Hidden Light: Science Secrets of the Bible

Dr. David Medved, Ph.D.

Hidden Light
Science Secrets of the Bible

Maggid Books

Hidden Light: Science Secrets of the Bible

First English Edition, 2008

Maggid Books

A division of *The* Toby Press LLC
POB 8531, New Milford, CT 06776–8531, USA
& POB 2455, London W1A 5WY, England
& POB 4044, Jerusalem 91040, Israel

The right of David Medved to be identified as the author of this work has been asserted by him in accordance with the Copyright, Designs & Patents Act 1988

All rights reserved. No part of this publication may be reproduced, stored in a retrieval system, or transmitted in any form or by any means, electronic, mechanical, photocopying or otherwise, without the prior permission of the publisher, except in the case of brief quotations embedded in critical articles or reviews.

ISNB 978-1-59264-185-7, *hardcover*

A CIP catalogue record for this title is available from the British Library

Typeset in Minion by Koren Publishing Services

Printed and bound in the United States

Contents

CHAPTER I: COSMOLOGY AND THE BIBLE 1

A. Introduction – Creation According to the Bible,
   Modern Science and Pagan Accounts 2
   1) Genesis – The Six Days of Creation 2
   Appendix 1.1 – Biographical Note on Moshe Dove Medved. . 7
   2) Other Creation Stories 8
B. Day 1 – How It All Began (The Origins of the Universe
   According to Science) 10
   1) Red Shift and the Expanding Universe. 10
   2) Steady State vs. The Big Bang 12
   Appendix 1.2 – Biographical Note on George Gamow ... 16
   Appendix 1.3 – Ryle vs. Hoyle 18
   3) The Cosmic Microwave Background Radiation (CMBR). 19
   4) The Light of Genesis 1, Chagiga 12a and the CMBR 23
   5) Edgar Allan Poe on Olbers' Paradox and the
      Big Bang, 100 Years Before Gamow 25
C. The Second Day............................... 27
   1) What is the Raqiya?.......................... 27
   2) Recent Research Results from High Energy
      Collision Experiments 29

3) Other Commentaries on the Raqiya 30
   4) Where Has All The Anti-Matter Gone? 31
   Appendix 1.4 – The Wave Equations of Schrödinger
      and Dirac: Matter & Anti-Matter 37
   Appendix 1.5 – Talmudic and Rabbinic Exegetes 42
D. Resolution of the Contradiction between the Chronology
   in Genesis and Cosmogony – for Days Three and Four. . . . 43
   1) The Nature of the Primordial Atmosphere 43
   2) Problems with Rashi's Commentary on Genesis 11:5–8 . 44
E. The Age of the Universe 46
   1) Does the "Age" Depend on Location? 46
   2) A 13th Century Calculation 47
   3) The Age of the Universe in Christian Thought 49
   4) An Alternative Calculation 50
F. The Fate of the Universe: Expansion Forever 51
   1) Philosophers and Poets 51
   2) The Friedmann Solutions to the Einstein Equations . . . 53
   3) Biblical Sources 55
   4) Science and Eschatology (the End of Days) 57
   Appendix 1.6 – Leviticus XXV and the
      Seven Sabbatical Cycles 63

**CHAPTER II: THE MUSIC OF THE CELESTIAL SPHERES
(ASTRONOMY AND THE BIBLE)** 67

   1) Psalm 19 (King David, ca. 1000 BCE) 68
   2) From Pythagoras (Sixth Century BCE) to Holst 70
   3) Bode's Law and Musical Octaves 73
   4) The Secrets of Science in Psalm 19 74
   5) Conclusions 82
   Appendix II.1 – God of Wonders 85
   Appendix II.2 – Timeline of the Concept 87

CHAPTER III: PI(Π) IN THE BIIBLE
(MATHEMATICS AND THE BIBLE) 89

   1) A brief history of Pi............................. 90
   2) Critique, Invective and Worse..................... 95
   3) Apologia On Behalf of the Biblical Value........... 97
   4) The Revealed Value of Pi in the Bible 97
   Appendix III.1 – Huram of Tyre 100

CHAPTER IV: CHEMISTRY AND THE BIBLE 103

A. Biblical Blue (TEKHELET): Origins and Chemistry 104
   1) Questions on the Book of Numbers xv:38–39. 104
   2) Commentaries on the Questions.................. 105
   3) Biblical Blue (tekhelet) and Tyrian Purple
      (argaman) – History and Sources................ 108
   4) The Chemical Structures 114
   5) The Physics of Color and Photochromic Processes.....118
   6) Tekhelet and the Flag of Israel 121
B. The Structure of Water and Its Amazing Properties 122
   1) The Water Molecule............................ 122
   2) Water in the Bible 124
   3) Concluding remarks 126
C. Other Forms of Water: Ice and Snow 126
   1) The Color of Ice................................ 126
   2) Snow in the Bible 127

CHAPTER V: LUNAR-SOLAR PERIODICITIES OF
LARGE EARTHQUAKES ON NORTH-SOUTH RIFTS......... 131

   1) History of Earthquake Predictions..................131
   2) The Kilston-Knopoff Paper on Lunar-Solar
      Periodicities of Large Earthquakes in
      Southern California..........................133
   3) Earthquakes in the Bible (The Jordan Rift) 139
   4) Large Earthquakes on the Jordan Rift since 31 BCE 145

5) The Great Indonesian Earthquake and Tsunami of
     December, 2004. 148
  6) Conclusions and Forecasts. 148
  List of Figures 149

CHAPTER VI: ARCHAEOLOGY AND THE BIBLE
(THE EXODUS FROM EGYPT) 151

  1) Open Questions on Exodus. 151
  2) The Date According to the Bible 152
  3) The Route According to the Bible 154
  4) Archaeological Evidence 157
  5) Dating Problems 159
  6) Is Evil (רעה) MARS?. 160

ACKNOWLEDGEMENTS 167

Synopsis

Over the past two decades, a number of books, conferences and articles have been devoted to demonstrating the harmony between Science and the Bible. Stimulated by advances in cosmology and biology, authors like Paul Davies, Gerald Schroeder and Nathan Aviezer have provided us with new insights into the compatibility between a scientific and religious world view. Organizations such as the Templeton Foundation, b'Or Hatorah and the Association of Orthodox Jewish Scientists have actively promoted the unity of the latest scientific theories with spirituality and Biblical exegesis. The Discovery Institute has a number of distinguished scientists developing a Theory of Intelligent Design. Much of this work has focused on some specific aspect of the Biblical narrative, especially the Book of Genesis and its connection with a particular scientific theory. For example, in his book "Genesis and the Big Bang" (Oct. 1990), Dr. Gerald L. Schroeder hypothesizes how the six days of creation are consistent with the 15.6 billion years of the Universe and he describes Nahmanides' idea that the entire Universe started from "a speck no larger than a mustard seed." In the same year, Prof. Nathan Aviezer published "In the Beginning" with a different viewpoint on the same theme. Dr. Schroeder subsequently wrote and published the "Science of God" and "The Hidden Face of God," and Prof. Aviezer has recently published "Fossils and Faith," which includes an evocative treatment of faith, prayer and miracles.

The Australian physicist Paul Davies has authored a number of related works such as "The Mind of God" and "God and the New Physics." Rabbi Aryeh Kaplan in his slim text, "Immortality, Resurrection and the Age of the Universe: A Kabbalistic View" (1993), attempts to resolve the conflict between science and the Bible, centered on the question of the age of our world. A book entitled "The Privileged Planet" (2004) by Guillermo Gonzalez and Jay W. Richards describes the unique position of Earth in the solar system and in our galaxy. Detailed scientific arguments by the authors present a serious challenge to the expositors of the Copernican Principle (that man is merely an impure lump of carbon crawling about on the surface of an insignificant speck of cosmic dust).

This present work explores and explains the heretofore "hidden secrets of science" embedded throughout the Biblical text. In this respect, it differs significantly from these earlier publications in scope and content. Instead of concentrating on a specific scientific discipline, this book spans six major fields from cosmology and astronomy through mathematics to chemistry, geology and archaeology. Similarly, the biblical sources are drawn from most of the 24 Books of the Bible, ranging from Genesis to Deuteronomy, Joshua and Judges, Kings I and II, Psalms, Chronicles, The Twelve Prophets and the Book of Daniel. Historical, literary and artistic sources are integrated with the biblical and scientific material. These authors and philosophers range from Pythagoras and Pliny, to Shakespeare, Edgar Allan Poe, Robert Frost and T.S. Eliot.

Our voyage of discovery proceeds without recourse to mysticism or biblical codes, while providing a unified treatment over a broad range of topics, revealing the mind-boggling depth and accuracy of the Bible.

Where appropriate, the latest scientific discoveries are cited. For example, the section on the age and fate of the universe contains conclusions based on observations from the Hubble Space

David Medved

Telescope, the COBE satellite and the most recent data from the WMAP satellite. Biblical sources on sabbatical cycles are compared with this data, leading to new insights on the models of the oscillating cosmos. The chapter on astronomy demonstrates how the glorious language of Psalm 19 anticipates the most recent findings in pulsars, binary star systems and the gravitational wave projects currently in progress.

The text should stimulate dialogue, vigorous debate and even collaboration between scientists and biblical scholars, encourage the teaching of science and classics in religious schools, and provide a user-friendly introduction to the Bible in high schools and universities. The target audience of our discussion includes Orthodox and Traditional Jews and a goodly fraction of Evangelical and Traditional Christians, as well as a wider base of the secular population seeking a rational connection to The Force.

Chapter 1
Cosmology and the Bible

This Chapter begins with an overview of the six days of creation in the Book of Genesis and develops comparisons and analogies between science and the Biblical account. In some respects, the approach is similar to the familiar Day-Age model used by other authors, but with significant differences in the details. In general, this Day-Age method posits that each day corresponds to a specific epoch in the evolution of the Universe. Contradictions and discrepancies are duly noted with their resolution deferred to later sections of the chapter. The conflict between the Big Bang and Steady-State models of the Universe is described in detail as part of our discussion on Day One. Each succeeding day is then covered in separate sections, up to Days Five and Six, which describe the appearance of life on Earth. The Second Day provides the greatest difficulty, as it has for many generations of scholars and commentators. The "raqiya," variously translated as firmament, dome, expanse, etc., is identified in Section C as the barrier between matter and anti-matter – a speculation based on analysis of Biblical language and commentaries. An apparent contradiction between the Third and Fourth Days is resolved in Section D by discussion of light scattering in the primordial atmosphere.

Hidden Light: Science Secrets of the Bible

This chapter concludes with a comparison of scientific and Biblical sources on the age and fate of the Universe, in Sections E and F.

A. INTRODUCTION – CREATION ACCORDING TO THE BIBLE, MODERN SCIENCE AND PAGAN ACCOUNTS

1) Genesis – The Six Days of Creation

The Bible starts with a simple seven-word sentence:
"בראשית ברא אלהים את השמים ואת הארץ".

"In the beginning God created the heaven and the earth"
   In this very first sentence, the Torah raises several questions, such as:

1. Why is the first letter written extra large (the big *Bet*)?
2. How could it say that heaven and earth were created at the "beginning" when the earth doesn't make its appearance until Day 3?

In response to question 1), note that the Big *Bet* opens on the left and is closed on the right. Hebrew is written from right to left. This oversized first letter of the Bible acts as a time barrier, with time's arrow flowing from right to left following the written language.
   We cannot penetrate to the right of "Time Zero" (as in the latest theoretical models of the Big Bang – "The Beginning"). The Bible is anticipating the difficulties encountered by our contemporary cosmologists, who are struggling with the singularities at Time Zero. It is interesting that this scientific term ('singularity'), in reference to the point of Time Zero, could be interpreted as descriptive of the Oneness and Unity of God.
   Professor Leon Kass has succinctly addressed this point in a recent issue of Commentary Magazine (Ref. 1.1). He writes:

> "In Cosmology, we have seen wonderful progress in characterizing the temporal beginnings as a big bang and elaborate calculations to characterize what happened next. But

from science we get complete silence regarding the status quo ant and the ultimate cause. Unlike a normally curious child, a cosmologist does not ask – what was before the big bang? – because the answer must be an exasperated 'God only knows.'"

One response to question 2) was provided by Rashi[1], who suggested that this first sentence should run together with the succeeding sentences such:

"At the beginning of the Creation of heaven and earth when the earth was without form and void and there was darkness on the face of the murmuring deep..." Some contemporary authorities are of the opinion that this first sentence encompasses the first ten billion years of creation which includes the formation of the stars, the solar system and the earth (private communication from Prof. Joseph Bodenheimer, President of the Jerusalem College of Technology).

This author prefers the following alternative interpretations of the first sentence:

1. In the beginning God created the spiritual world (*Shomayim*) and the physical world (*Eretz*). These are two orthogonal, non-intersecting domains which only meet under very special conditions.
2. In the beginning God created space-time (*Shomayim*) and compacted matter (*Eretz*).

Continuing with the Biblical narrative (verse 3): "And God said let there be light and there was light...and it was evening and it was morning, one day (יום אחד)." Once again we are confronted with an apparent chronological contradiction as in Question (2) on the opening sentence. The Bible speaks of evening and morning and yet the rotating Earth hasn't even been formed. There is also the related question, what is meant by one day?

---

1. *Rashi* (Rabbi Shlomo Itzhaki), b. 1040, France

The description of the second day is even more mystifying (verse 6):

> "And God said, let there be a *raqiya* (רקיע) *in the midst* of the waters and let it cause a division between waters and the waters. The "raqiya" has been variously translated as a dome, firmament or expanse. Nachmanides (Ramban) has noted that these verses touch upon the innermost mysteries of Creation and states: "Do not expect me to write anything about it" (Ref. 1.2). With the hindsight of modern science, a speculative solution to the mystery of the Second Day is proposed in Section C of this chapter.

The Biblical narrative continues to describe the formation of our Earth on the Third Day and the placement or appearance of sun, moon and stars on the Fourth Day. Modern cosmology holds that the stars and our sun were formed well before the solar system and Earth. This apparent contradiction will be addressed in detail in Section D of this Chapter.

The description of The Fifth and Sixth Days starts with life in the seas and the formation of higher order species, culminating in the creation of Adam (first man) on the evening of the Sixth Day. There is a remarkable similarity between the chronology of the Biblical account and the current theories of cosmology, cosmogony[2] and evolution. As shown in Table 1 (page 5), there are two cases where the chronologies do not seem to track. In the first case, the transition from the end of Day I to the formation of the mysterious Raqiya on Day II seems to have a time slip of 370,000 years (column 4 of the Table). A possible solution is discussed in detail in Section C. The apparent discrepancy between the Third and Fourth Days is resolved in Section D of this chapter.

---

2. Cosmogony is the study of the solar system and its origins, whereas cosmology refers to the study of the universe.

David Medved

Table 1: Comparison of Six Days of Creation with Modern Science*

verse	Biblical Account	Modern Science	Estimated elapsed time after the Big Bang
1a 1b	Creation of matter, energy, space, time First Light	The Big Bang – Dark Ages of the Universe CMBR (Cosmic Microwave Background Radiation)	The first three minutes According to the Standard Model 370,000 years
2a 2b	*Rakia* Cycles of dark and light (erev v'boker) Commentaries of the sages	Separation of matter from anti-matter Speculative model that the First Light were high-energy gamma rays down-shifted to today's soft x-ray background (section c) Stars and sun	Right after the Big Bang (about 5 seconds) End of the recombination epoch at 5 seconds 200 million years to 8 billion years
3	Earth, plants, oceans, seas	Earth and moon, our solar system	9 billion years
4	Sun, moon and stars	Transformation of Earth's atmosphere	10 to 11 billion years
5	Life begins in seas, fish, reptiles, birds	Life begins in seas, fish, reptiles, birds	13 billion years prox.
6	Land mammals Adam on the evening of the sixth day	Mammals, hominids Man	13.6 billion years (100 million years ago) Open Question – biblical dating says 5768 years ago

* based on the age of the Universe at 13.7 billion years, as determined by data from the WMAP satellite – see Section E.

As noted in Table 1, the biblical chronology of the Fifth and Sixth Days dealing with the appearance of life and the creation of higher-order species gives astonishingly similar results to the Theory of Evolution, except for one fundamental difference. The Bible implies that each species was independently created (with evolution possibly occurring within a given species), whereas Darwinian theory talks about an evolutionary ladder between each species, culminating in Man. Evolutionists point out the amazing similarities in the DNA chains between Man and chimpanzees which are cited as proof of a common ancestor. There is a difference of only two percent in the genetic sequences. However, one searches in vain for some explanation on how such a small difference can account for the miniscule intelligence of a species more than one million years old, compared to their human "cousins." Also, until such time as there is clear evidence from the fossil record showing the transition phases from one species to the next, each model should be given equal consideration and taught in our schools and universities in the spirit of free enquiry.

David Medved

Appendix 1.1 Biographical Note on Moshe Dove Medved
One of my earliest memories as a young child growing up in South Philadelphia was the emblem painted on the sides of my brother's truck. My brother, Moshe Leib Dove was a struggling electrical contractor during the dark days of the Great Depression. Written in both Hebrew and English were the words יהי אור (Let there be Light) together with three lightning bolts originating from a common point. This message from Genesis resonated so well with the good citizens of Greater Philadelphia that Dove Electric became a household word. Although he was not religiously observant, his self-taught knowledge of the Bible and his love of Zion were transmitted to me and subsequently to my sons when he came to live with us in the Santa Monica mountains in the last ten years of his life. His influence on our family as Uncle Moish has been eloquently described in my son Michael Medved's recent book "Right Turns."

2) Other Creation Stories

Generations of secular scholars have tried to draw parallels between Genesis and ancient Middle Eastern accounts of the origins of the world. These parallels exist only in their imaginations, since in none of these graphic, even obscene pagan accounts, is there any anticipation or prediction of the development of the Universe as described by modern cosmology. Rob Yule, previously at St. Albans Presbyterian Church (in New Zealand), has written: "Genesis 1… is strikingly different from the other creation accounts…yet it anticipates many of the most remarkable findings of modern science" (Ref. 1.3).

The Egyptian papyrus of Apophis describes how their sun god (Ra) created the world and mankind by using his salivating mouth to spit out progeny. A less sanitized version deals with masturbation and autofellatio. Apophis is a seven-headed monstrous serpent who is in constant warfare with the sun god. In the Babylonian epic of Enuma Elish, we read about a drunken Marduk, who after killing and dismembering Tiamat, performs acts of magic in front of the gods and then proceeds to create the world while killing off his competitors. Their gods are fickle, bribable, malicious, irrational and capricious. These and other distasteful depictions of their deities are cultural ancestors of today's R- or X-rated movies. One could write a great Ph.D. thesis tracing the connection between these 4000-year-old pagan stories and the current Hollywood ethos. Both Marduk and Apophis (mistakenly called Aphosis) are names prominently featured in the popular media, from video games to television. In the TV series Stargate SG1, Marduk was a Goa'uld who ruled over a planet with great brutality. His own priests revolted and sealed him in a sarcophagus. In the same series, the Goa'uld Death Glider was designed and built by Aphosis. This fascination by the lords of the entertainment industry with the names of these ancient deities is a clear indication of the connection.

At least the Gilgamesh Epic, which has often been cited as an "inspiration" for the story of Noah and the Flood, does have some redeeming PG features. The Epic is an ancient Babylonian poem about a mythological hero-king thought to have ruled during the Third Millennium BCE (around 2600 BCE) – corresponding to the year 1167 on the Hebrew calendar. According to tradition, the Great Flood which covered the whole earth occurred in 1656 when Noah was 600 years old. By contrast, the Babylonian flood myth describes a large deluge on a local river.

On the continent of Australia, halfway around the globe from the Middle East, the nomadic Aborigines developed stories of creation which have more in common with the Biblical account than those of Israel's neighboring cultures. Consider these characteristics:

1. Although their depictions vary from tribe to tribe, all have the oncept of the "Dreamtime" – the time which existed before the creation of the world (cf. our discussion of the Big *Bet* in Section A.1).
2. It was dark and there was no space between earth and sky, the magpies had no space to fly.
3. With a great concerted effort, they began to lift the dome of the sky with their beaks.
4. As the magpies watched, the sky split open and the sun appeared.
5. The magpies (and other birds) have never forgotten that first glimpse of the sunlight and consequently greet the sunrise every day with a joyous chorus of song.

Contrast this lyrical vision of creation from a primitive culture with the degeneracy of those advanced civilizations so much admired by the elitist literati of the Western world.

At the start of every day, observant Jews praise the handiwork of the Creator with prayer and joyous song, including the Hallelujahs of Psalms 145 to 150, culminating in the words of Psalm 150 – "Every soul will praise Thee Hallelujah." Many Christians, including President Bush, also begin their day with inspirational readings from the Bible.

B. DAY 1 – HOW IT ALL BEGAN (THE ORIGINS OF THE UNIVERSE ACCORDING TO SCIENCE)

1) Red Shift and the Expanding Universe

The competing models of the Big Bang (BB) and the Steady State Universe (SSU) were both attempts to explain the expansion of the Universe as observed in the several decades before 1948. As early as 1912, Vesto Slipher, working with Percival Lowell in Arizona, was able to obtain spectra of spiral nebulae with a red shift, corresponding to velocities of recession[3] up to 300 km/sec (one thousandth the velocity of light). This speed was considered to be so large that astronomers questioned the data.

In 1922–24, Alexander Friedmann, a Russian-born meteorologist and mathematician, published his three solutions of the Einstein Equations of General Relativity. As a firm believer in a static universe, Einstein himself had introduced the Cosmological Constant, which became known as one of the greatest fudge factors of science.

In fact, Einstein introduced this concept into his Equations of General Relativity, as some mysterious, repulsive energy which exactly balanced the gravitational contraction of matter in the Universe. The idea of a static universe goes back to the Greek philosophers and the Hindus. When confronted by the evidence of an expanding universe, he declared that his introduction of the cosmological constant was "...the biggest blunder of my life." About 70 years later (in the late 1990s), it was found that this may not have been Einstein's "biggest blunder", but perhaps his greatest legacy (see section F).

The Friedmann solutions predicted an expanding universe, and they also predicted the beginning and the end of Time. In Sections E and F (on the Age and Fate of the Universe), these three solutions will be discussed in detail.

---

3. Actually, out of 25 nebulae analyzed, 3 (including Andromeda) seemed to be approaching Earth. This anomaly was later explained as a large random velocity superimposed on the smaller expansion velocity.

In 1927, Georges Lemaître, a Belgian Catholic priest working in Astrophysics, published his model of an expanding universe, which was completely overlooked until 1931 when Einstein and de Sitter (a Dutch cosmologist who also believed in a static universe) publicly praised his work.

Beginning in 1928, Edwin Hubble and his assistant, Milton Humason, at the Mt. Wilson Observatory in California, began a systematic study of red-shifted spectra from faint distant nebulae and discovered a linear relationship between their distance (r) and their velocity (v) as given in

$$\text{Eq. 1: } v = H_0 r.$$

H_0 is known today as the Hubble constant. An updated extension of the original results is shown in Fig. I.1. The region surveyed by Hubble and Humason as shown near the origin is roughly 10% of the current data. Hubble's constant has dimensions of reciprocal time.

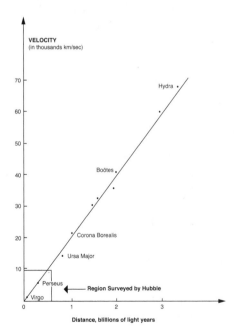

Fig I.1 The Hubble Diagram Galaxy Recession Velocity as a Function of Distance

Hidden Light: Science Secrets of the Bible

One Light Year (LY) is the distance traveled by light in one year and is readily computed as

300,000 $\frac{km}{sec}$ (speed of light) × 31,557,600 (no. of seconds in one year) = 9.467×10^{12} km or approximately 10^{13} km. One billion light years is 10^{22} km., so the Hubble constant is 2×10^{-18} sec^{-1}. The convergence of Relativity, Quantum Mechanics and the model of the Expanding Universe just before World War II became the basis for the New Astrophysics of our post-war world.

2) Steady State vs. The Big Bang

The theory of the Steady State Universe (SSU) was developed around 1950 by Fred Hoyle, Hermann Bondi and Tom Gold as an alternative to the early Theory of the Big Bang (BB), first propounded by George Gamow in 1948. In fact, Hoyle invented the term, intending it as a pejorative (there never had been a "big bang" according to Hoyle). By 1948, most scientists had accepted the model of the expanding universe based on the observations of the red shift by Edwin Hubble. Hoyle, et al., proposed that as the galaxies moved apart, new galaxies were formed out of spontaneously created "new matter" (about one hydrogen atom every four cubic kilometers). These new galaxies filled up the voids produced by the expansion of the Universe. As a result, the Universe would look the same in the future as in the past. The density of matter and energy would stay approximately constant.

Many scientists (especially in England) favored the SSU model over the BB theory since the latter idea of a beginning to the Universe implied Divine Intervention. A much smaller number of religious scientists preferred the BB model for that very reason. Many philosophers and scientists have compared the SSU to the Aristotelian view of the world and also to concepts in the Hindu and other Far Eastern theologies. The controversy raged on for nearly two decades, punctuated by alternating moments of bitter invective and hilarious humor. Hoyle on the BBC was quoted as saying: "This instantaneous creation of the Universe is like a party girl jumping out of a cake." George Gamow and his wife, Barbara, composed verses during the early 1950s, poking fun at Hoyle and his supporters (see Appendix 1.2 on George Gamow).

The Universe...

Gamow imagined Hoyle in a cosmic opera, suddenly materialized from nothing in the space between the brightly shining galaxies, bursting majestically into song:

The universe, by Heaven's decree
Was never formed in time gone by,
But is, has been, shall ever be –
For so say Bondi, Gold and I.
Stay, O Cosmos, O Cosmos, stay the same!
We the Steady State proclaim!

The aging galaxies disperse,
Burn out, and exit from the scene.
But all the while, the universe
Is, was, shall ever be, has been.
Stay, O Cosmos, O Cosmos, stay the same!
We the Steady State proclaim!

And still new galaxies condense
From nothing, as they did before.
(Lemaître and Gamow, no offence!)
All was, will be for evermore.
Stay, O Cosmos, O Cosmos, stay the same!
We the Steady State proclaim!

From 1948 to about 1960, the SSU had a major advantage over the BB. When the rate of expansion was extrapolated backward it would result in an age of the Universe much less than the age of the older stars. Later it was shown by Walter Baade that the size of the Universe had been underestimated by a factor of two and this resulted in an age of the Universe significantly greater than that of the solar system and the stars.

By 1955, Martin Ryle at Cambridge had surveyed almost 2,000 radio stars and determined that most of them were outside our galaxy. It took another nine years of hard work, using better radio telescopes, for his results to be accepted. Hoyle complained with some bitterness that Ryle was motivated by some deep desire to destroy the SSU, whereas Ryle commented that "…(theoretical) cosmologists…have always lived in a happy state of being able to postulate theories which had no chance of being disproved." There was no love lost between observational astronomers and theoretical cosmologists. Lev Landau, a great physicist of the Golden Era in Physics (1930–1950), once commented that "cosmologists are seldom correct, but have no doubt that they are always right."

Barbara Gamow wrote a long ditty following the Hoyle-Ryle row (only two stanzas are given here. See Appendix 1.3 for the full poem):

Ryle vs. Hoyle
Commentary on Ryle versus Hoyle by Barbara Gamow, George Gamow's wife

"Your years of toil,"
Said Ryle to Hoyle,
"Are wasted years, believe me.
The steady state
Is out of date.
Unless my eyes deceive me,

My telescope
Has dashed your hope;
Your tenets are refuted.
Let me be terse:

Our universe
Grows daily more diluted!"

Ryle's radio stars and the publication of the discovery by Penzias and Wilson of the Cosmic Background Radiation in 1965 heralded the demise of the SSU, except for a few die-hards who invented the quasi-SSU. This feeble innovation proposed that the new matter came into being by a series of mini big bangs. There may still be some scientists who support this model and its variations.

Appendix 1.2
Biographical Note on George Gamow

George Gamow was an Ukranian-born physicist who joined the faculty at George Washington University after coming to the U.S. in 1934. He had two brilliant collaborators – Ralph Alpher, his graduate student at GWU, and Robert Herman, who worked in the Applied Physics Lab. Alpher and Herman had predicted that a background remnant of the primordial fireball from the Big Bang should still be detectable.

A famous paper by Alpher and Gamow on cosmogenesis and the abundance of elements in the early Universe became known (perhaps notoriously) as the Alphabet Paper.

Not one to pass up the opportunity for a good joke, Gamow had sent the manuscript to Hans Bethe and asked the Nobel Laureate's permission to insert his name as a co-author on the work (even though Bethe had nothing whatsoever to do with this work). The paper was accepted and duly published in the Physical Review, the premium journal of the physics community, with authors Alpher, Bethe and Gamow. Rumor has it that Herman refused to change his name to Delter and missed out on his chance for fame and glory.

In 1956, before joining the University of Colorado, George Gamow came to San Diego, California at the invitation of John Jay Hopkins, a visionary entrepreneur and president of General Dynamics. Prof. Gamow gave a series of lectures to a large mixed audience of old-line airplane designers, aerospace engineers and physicists like this author (who had joined the Convair division of General Dynamics just after receiving his Ph.D. a year earlier). We became good friends and he was a frequent visitor at our home, where my dear wife of blessed memory would be constantly amazed at the prodigious amounts of vodka he was able to consume without any visible effect.

On one memorable evening he decided to teach me and my good friend, Dr. Wade Fite, "how to hear music." He demonstrated by taking a short broomstick, placing the palm of his hand on the top and bending over so that his head rested on the back of his hand. We were then instructed to pivot quickly around the broomstick five times and then stand up straight, at which point we would "hear music." The effect was quite dramatic. Wade and I both went sprawling across the floor.

It was only a few years later, while engaged in qualifying for the NASA Scientist-Astronaut Program[4], that I learned about nystygmus and gyroscopic fluids in the head. I also had the privilege of learning from George Gamow, in private conversations, about his views on the new satellite age just dawning, and I reciprocated by attempting to teach him about the physics of transistors. We discussed the possibility that there may be other planetary systems around the stars and how one might overcome the scintillating effects of the Earth's atmosphere to observe such bodies. The concept of a space telescope in orbit above the atmosphere was conceived around that time (some forty years before the launch of the Hubble space telescope).

---

4. The Scientist-Astronaut program was initiated by NASA in the early 1960s to select Ph.D. scientists and MDs as active participants for future missions to the moon and beyond.

Appendix 1.3
Ryle vs. Hoyle
by Barbara Gamow, George Gamow's wife:

"Your years of toil,"
Said Ryle to Hoyle,
"Are wasted years, believe me.
The steady state
Is out of date.
Unless my eyes deceive me,

My telescope
Has dashed your hope;
Your tenets are refuted.
Let me be terse:
Our universe
Grows daily more diluted!"

Said Hoyle, "You quote
Lemaître, I note,
And Gamow. Well, forget them!
That errant gang
And their Big Bang –
Why aid them and abet them?

You see, my friend,
It has no end
And there was no beginning,
As Bondi, Gold,
And I will hold
Until our hair is thinning!"

"Not so!" cried Ryle
With rising bile
And straining at the tether;
"Far galaxies

Are, as one sees,
More tightly packed together!"

"You make me boil!"
Exploded Hoyle,
His statement rearranging;
"New matter's born
Each night and morn.
The picture is unchanging!"

"Come off it, Hoyle!
I aim to foil
You yet" (The fun commences)
"And in a while"
Continued Ryle,
"I'll bring you to your senses!"

3) The Cosmic Microwave Background Radiation (CMBR)

In 1964, Arno Penzias and Robert Wilson, two radio astronomers at the Holmdel Facilities of Bell Telephone Laboratories, were engaged in the difficult work of measuring the absolute source strengths of radio stars. They were troubled by an unexpected noise or hiss which added to the noise of their electronic circuits. They even went to the trouble of chasing out the pigeons who had made a comfortable home in their horn antenna (15 meters in length and 4×4 meters at the aperture) and cleaning out the accumulated guano (which Penzias euphemistically described as a "white dielectric substance"), but to no avail. No matter which way they pointed their highly directional antenna, the noise persisted with essentially equal amplitude from all directions, i.e., it was coming from an isotropic source at the receiver frequency of their radio telescope (approx. at 4 Gigahertz).

It turns out that Holmdel is not far from Princeton University, where R.H. Dicke and his colleagues were cognizant of Gamow's pioneering work on the Big Bang. Alpher and Herman had published a paper which predicted that a remnant of the radiation from the "primordial fireball" should still be detectable and they also gave an estimate of its temperature (Ref. 1.4). In a telephone call from Penzias, Dicke recognized the annoying hiss as a message from "the edge of the universe." He and his group were in fact preparing their own equipment to search for this radiation. As he hung up the telephone, he turned to his colleagues and uttered those words which are the nightmare of every scientist – "Boys, we've been scooped!"

In 1965 both groups published consecutive papers in the Astrophysics Journal on the theoretical interpretation and the experimental results (Ref.s 1.5 and 1.6), but only Penzias and Wilson received the Nobel Prize. As one reviewer put it, "The Big Bang has delivered a knock-out punch to its Steady-State competitor."

Subsequent measurements using satellites have confirmed that the CMBR is indeed a red-shifted remnant (by a factor of 1,000) of the Big Bang. Today's Standard Model of the Big Bang requires that the CMBR be isotropic (uniform in all directions) and have a thermal radiation spectrum. Fig. 1.2a shows the microwave images recorded and transmitted by these satellites.

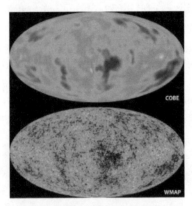

Fig 1.2a – the CMBR as seen by COBE and WMAP satellites. Note the improved resolution obtained by WMAP.

When the results from COBE (Cosmic Background Explorer) were first presented, they received a standing ovation from the audience of scientists and engineers. The remarkable agreement of the data points with the theory of black-body radiation is shown in Fig. 1.2b.

After the initial moments of the Big Bang, light was trapped inside the stew of ions and free electrons and was only able "to escape" when the temperature dropped below 3000° K, allowing atoms to be formed.

Free electrons will absorb photons, as will atoms, by a process known as Compton scattering (if the photons are energetic enough to ionize such atoms). At temperatures above 3000° K, corresponding to 0.25 electron-volts, there are enough energetic photons in the tail of the Maxwell-Boltzman distribution to keep free electron production in equilibrium with hydrogen atom formation. The equivalence between the temperature of a gas or plasma and the kinetic energy of its constituent particles will be discussed in more detail in Section 1C in connection with the temporal evolution of the Universe.

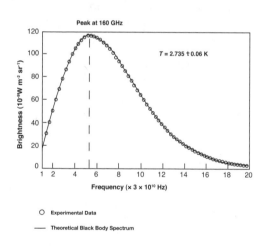

Fig I.2b – Radiation Spectrum of the CMBR as measured by COBE instrumentation

About 370,000 years after the Big Bang, as the Universe inflated and expanded, the temperature had dropped from values like 10^{12} degrees K to 3000°, at which point the trapped light could be emitted.

The CMBR is 1,000 times cooler today, corresponding to a Black Body Radiation temperature of approx. three degrees Kelvin (see Fig. 1.2b).

Two satellites have provided significant confirmations of this model. In 1989, NASA launched the COBE mission. It found that:

1. The temperature of the Universe is 2.735° K, which agrees with the predictions of the Big Bang.
2. The anisotropy (variation from uniformity in different directions) in the CMBR was found to be quite small, less than one part in 10,000. However, even such a small variation caused great excitement in the cosmological community, as it provided a view on what happened as matter coalesced in non-uniform clumps to form galaxies and clusters (as shown in Fig. 1.2a). The conditions under which the galaxies were allowed to form are known as the "Goldilocks Conditions," where the fluctuations in the ratio of temperature to matter must be "just right." If the fluctuations in this ratio were too large, the Universe would only consist of lots of massive black holes – if too small, there would be no stellar formation.

If the early Universe had been perfectly uniform, Days Three, Four, Five and Six would not have happened (no galaxies, no sun, and no earth).

In 2001, the WMAP (*Wilkinson Microwave Anisotropy Probe*) was launched with the capability of measuring temperature differences in the CMBR to a millionth of a degree. Wilkinson was a co-author with Dicke on the theoretical interpretation of the CMBR first detected by Penzias and Wilson. Sadly, he died before the data from the WMAP became available. The WMAP results are still undergoing analysis, but have already given us an "updated" value of the Age of the Universe at 13.7 billion years. These initial results also indicate that the first stars were formed 200 million years after the Big Bang, much earlier than previously thought.

More exciting results on "The Oldest Light in the Universe" will be discussed in greater detail in Chapter 11. The composite of

Fig. 1.2a was prepared from NASA data by Senior Minister Rob Yule of Greyfriars Presbyterian Church (formerly of St. Albans Presbyterian Church) in Auckland, New Zealand (see his recent booklet entitled "The Discovery of the Beginning" (2006), available from Affirm Publications, Tauranga, New Zealand).

4) The Light of Genesis 1, Chagiga 12a and the CMBR

The Torah Sages (first and second centuries) held conflicting opinions on the nature of the Light that was created on the First Day. Rabbi Elazar said, "The Light the Lord created on the First Day enabled Man to see from one end of the World to the other." Rabbi Elazar ben Shamua lived a long and productive life during the era of the Mishnah (second century CE). He was one of the Ten Martyrs, tortured and executed by the Romans for teaching Torah. An allegorical poem on these events is read during *Musaf* on Yom Kippur and in modified form on Tisha B'Av.

Some skeptics might say that the good rabbi, despite his great learning and intelligence, was a member of The Flat Earth Society. Not so! He based his opinion on the Hebrew word for world, *olam,* which can also be translated as Universe or infinity of space and time.

This relativistic concept was expounded by the Maharal of Prague at the end of the sixteenth century. About 25 years ago, my teacher, Rabbi Daniel Lapin, pointed out that in his opinion the Maharal had anticipated Einstein by some 400 years. With the passage of time I had forgotten the reference and was unable to respond to my editor's challenge to elaborate on this assertion. The Rabbi's son, Aryeh Lapin, was attending Yeshiva in Israel (June, 2007) and had become an honorary member of the Medved family often joining us for Sabbath meals. I asked if he could elaborate on his father's comments; five minutes later he came up with a detailed answer – see Ref. 1.7.

It is tempting to identify the relative uniformity of the CMBR as the remnant of that Light. Rabbi Elazar goes on to state that this light was subsequently hidden and "saved for the righteous." My grandson Itamar recalls reading that one reason this Light was hidden was its high energy/intensity, but we have been unable to locate the specific reference. In Section 1.C we discuss the problem of identifying the 25th word of Genesis (light) as the CMBR, which would lead to serious chronological contradictions. In today's Universe, the CMBR spectrum is that of a black body at a temperature of 2.7 degrees Kelvin. As shown in Fig. 1.2b, it has a radiation peak at 120 GHz., corresponding to 2.5 mm. Note that the observations of Penzias and Wilson were at 4 GHz., where the brightness of the radiation was about ⅕ that of the peak frequency detected by a satellite high above the earth's atmosphere. Since this spectrum was downshifted by a factor of 1000 from the original emission, the first light would peak at 2.5 microns in the near-infrared with a significant visible light component.

Rabbi Elazar continues with the statement, "This issue is debated by the Tannaim (the Torah Sages)…They say that this light originated from the 'meorot' (luminaries) that were created on the First Day but which were only placed in the heavens on the Fourth Day."

The original light of the Big Bang was trapped in the primordial plasma stew until the Universe expanded and the temperature dropped. Genesis 1:4 states "…and God separated between the Light and the Darkness." Darkness is characterized as a specific creation – not just the absence of light. In Isaiah XLV:7 we read, "He who forms the light and creates the darkness." In that sense, until Light and Darkness were separated, they were mixed together in that hot soup of quarks and gluons, matter and anti-matter, and finally photons and electrons. The CMBR is therefore the earliest direct evidence available to us of the evolution of our bouncing baby Universe. For earlier times (known as the Dark Ages of the Universe),we must rely on data obtained using high energy particle accelerators at facilities like Brookhaven, CERN and Stanford, and on recent studies of the BAO (Baryon Acoustic Oscillations), to be discussed in Chapter IIA.

David Medved

5) Edgar Allan Poe on Olbers' Paradox and the Big Bang, 100 Years Before Gamow

In 1826, German astronomer Heinrich Wilhelm Olbers described the "dark night sky paradox," following earlier comments by Kepler, Halley and Cheseaux that in a static universe the night sky should be as bright as the sun. The prevalent opinion of the time held that the universe was static, of infinite age and containing an infinite number of uniformly distributed luminous stars. Under such circumstances, every line of sight from the surface of planet earth would terminate in a star, resulting in a night sky of blinding bright light. An early attempt to solve this paradox used the argument that the universe is not transparent, i.e. the light from distant stars may be blocked or absorbed by interstellar dust, gas or opaque objects. This mechanism was discounted when it was pointed out that any matter blocking the star's light would undergo a rise in its temperature and reradiate the absorbed energy, possibly at longer wavelengths.

In 1848, Edgar Allan Poe proposed in a 55 page prose poem that the Universe is not infinitely old and that it is expanding (Ref. 1.8). Either idea resolves Olbers' Paradox. Even more remarkable is his model of the Big Bang in the same paper. He describes a primordial particle of Divine Origin, which radiated an immense number of minute atoms into empty space, whose tendency to coalesce and reunite (under their mutual gravitational attraction) is opposed by the original "diffusive energy" imparted at the Beginning. Poe also hypothesizes that our galaxy is part of a cluster and a belt of clusters and then proceeds to introduce the concept of multi-universes. He also employs the term "universe" on two levels – the universe of stars, and all things material and spiritual (see the discussion on Genesis 1.1 in Section I.A.1).

These seminal ideas put forth by Poe one year before his untimely death and 100 years before the work of Gamow, Alpher and Herman, were ignored and forgotten by the scientific community. With incredible prescience, he writes in the poem's preface, "What I here propound is true – therefore it cannot die – or if by any means it be now trodden down so that it die, it will rise again to the Life Everlasting."

Poe's views of most scientists are summarized in a futuristic allegorical passage (below), which could certainly be applied to today's practitioners of scientism and Darwinian evolution. Scientism is defined as "...an exaggerated trust in the efficacy of the methods of natural science applied in all areas of investigation" (Merriam-Webster Collegiate Dictionary – 1997).

"From a letter found in a bottle floating on Mare Tenebrarum (Sea of Shadows) in the year 2848 [1000 years in Poe's future] – the persons thus suddenly elevated by the Hog-ian philosophy [I guess he means Baconian] into a status for which they are unfitted – thus transferred from the sculleries into the parlors of Science – from its pantries into its pulpits – than these individuals a more intolerant set of bigots and tyrants never existed on the face of the earth."

C. THE SECOND DAY

1) What is the Raqiya?

The same problem we encountered in the first sentence of the Bible confronts us again on the Second Day:

ויאמר אלהים יהי רקיע בתוך המים ויהי מבדיל בין מים למים: ויעש אלוהים את הרקיע ויבדל בין המים אשר מתחת לרקיע ובין המים אשר מעל לרקיע ויהי־כן: ויקרא אלהים לרקיע שמים ויהי־ערב ויהי־בקר יום שני

The English translation follows:

> "And God said, Let there be an expanse [or firmament] *in the midst of the waters* and let it cause a division between waters and the waters. And God made the expanse and caused a division between the waters which were below the expanse and the waters which were above the expanse and it was so. And God called the expanse Heaven and it was evening and morning, a Second Day."

The analogous question to Day One is: How could it refer to waters 'above' and 'below' when the earth, its seas, its clouds, etc. only make their appearance on the Third Day? The standard commentary describes the waters 'below' as the oceans which completely covered the face of the young earth, and the waters 'above' as the clouds. More recently, with a superficial knowledge of modern science, some identify the waters above as the dirty ice which makes up the bulk of comets. Even more far-fetched is the proposition that some astronomers have identified water molecules in other galaxies.

There are manifold insights provided by the Commentators on this question, especially by the great Rashi, whose comments (although based on 12th century knowledge) offer incredible clues to the phrases in this passage. Consider "in the midst of the waters." According to Rashi: "In the exact center of the waters, because there is the same distance between the upper waters and the firmament (expanse) as there is between the firmament and the waters that are upon the earth."

About 20 years ago, I was struck by the similarity of this language to the concept of the forbidden energy gap which separates the states occupied by matter and anti-matter (see Appendix 1.4 for a semi-technical explanation of the Dirac relativistic wave equation and its solutions). Was it possible that the Bible was describing the Raqiya as the energy gap between the positive energy states and particles 'above' and those negative energy states and anti-particles 'below'? This question lay dormant until I commenced the writing of this Chapter.

As noted in Appendix 1.4, these negative energy states extend all the way down to negative infinity. I am reminded of the distinguished lady who gave a lecture on the ancient Hindu concept that the world is supported on the back of a giant turtle. When asked in all seriousness by a student in the audience, "Dear Madam, on what does this turtle stand?" she sternly responded, "Young man, it's turtles all the way down!" Another version of the story identifies the skeptical student as a young Bertrand Russell (1872–1970), who became a famous philosopher, logician, mathematician and political activist and iconic idol to the young liberals of the 20th century for his espousals of socialism, free love, and trial marriage. We shall encounter him again in Section 1.E on The Fate of the Universe.

There are actually three levels of speculation on the meaning of the raqiya and its connection to the forbidden gap in the Dirac energy level solutions, exhibited in Fig. 1.3 of Appendix 1.4:

I. it's the barrier between the permitted positive energy and negative energy states;
II. it's the separation in energy between matter and anti-matter in our Universe;
III. it's a permanent barrier which prevents the mutual annihilation of all matter with anti-matter.

There are some serious questions about these speculations:

A. The word for waters is מים (*mayim*). It is a bit of a stretch to try to identify the primordial plasma as such.
B. A serious problem is in the last sentence, "And God called the expanse Heaven" ("ויקרא אלוהים לרקיע שמים...").
C. The chronology doesn't fit the current models of the early Universe, if we wish to maintain the Day-Age thesis of Table 1.1 in Section I.A.

In the following subsections, each of these questions will be addressed in detail.

2) Recent Research Results from High Energy Collision Experiments (as a possible response to question a.)

There was a report given at a 2005 meeting of the American Physical Society, describing experiments at Brookhaven National Laboratory, using the Relativistic Heavy Ion Collider (RHIC) (Ref. 1.9). The energy of the collisions was high enough to simulate conditions at 10-8 seconds after the Big Bang with temperatures in excess of 1012 degrees Kelvin. In their report, Sam Aronson and his colleagues conjured up an image of the quarks and gluons behaving as "a very nearly perfect liquid." To quote from their report:

> "When physicists talk about a perfect liquid, they don't mean the best glass of champagne they ever tasted. The word "perfect" refers to the liquid's viscosity, a friction-like property that affects a liquid's ability to flow and the resistance to objects trying to swim through it. Honey has a high viscosity; water's viscosity is low. A perfect liquid has no viscosity at all, which is impossible in reality but useful for theoretical discussions."

The Torah was written in the simple language of human beings on the surface of our planet 3,300 years ago, but still contains much deeper meanings for people in the 21st century. From the standpoint of a hypothetical observer, the liquid-like Early Universe may have been simply described as *mayim*.

The prescience of Rashi continues to astonish. In his commentary on verse 1.6 ("Let there be an expanse"), he writes as follows: "Let the expansion become fixed, for although the heavens were created on the first day, they were still in a fluid state and they became solidified on the second day...."

3) Other Commentaries on the Raqiya (in response to question b.) Ibn Ezra states that *Raqiya* means something that is stretched out (see Isaiah, XLII:5). We also encounter the same root in Exodus XXXIX:3 with reference to the Ephod (vest) of Aaron ("וירקעו את פחי הזהב"): "...and they did extend (stretch) the gold plates." Rashi comments on this line in connection with line 6 of Psalms CXXXVI: "To Him who stretched out the Earth over the waters for His kindness endures forever." This Psalm, known as the Great *Hallel*, is recited every Sabbath morning and also at the Passover *Seder*. It outlines the primal events of the Creation of the Universe and details the major happenings in the Exodus of the Israelites from Egypt. There are 26 verses corresponding to the numerical value of the Tetragrammaton. *Pirkei d'Rabbi Eliezer* states that this firmament is *not the same* as the heaven of Day One (see App. 1.5 for more detail on this and other Rabbinic sources). The Midrash picks up on this and says that were it not for the *Raqiya*, the Earth would be engulfed in a cataclysmic reaction between the waters above and the waters below. This is certainly evocative of matter and anti-matter galactic collisions, which have not been observed to date (see Section C.4 following).

B'chor Shor states that the word *mavdil* (ויבדל בין המים) denotes a permanent division. Some commentators state unequivocally that calling the Raqiya "Heaven" requires profound investigation. Rashi proposes in his comment on verse 8 ("and God called the firmament Heaven") that the word used for Heaven – שמים (*shamayim*) – is a compound, in this case, of two words: שם + מים (water and there), meaning "there is water there." Rambam states the word *shamayim* designates the whole extra-terrestrial universe surrounding the earth, whereas *Raqiya* designates that point in our atmosphere demarcating the beginning of the heavenly realm where human beings cannot survive (obviously written before the age of space travel).

Many other sources have commented on Verse 8, but it is the opinion of this writer that none has provided any better explanations.

4) Where Has All The Anti-Matter Gone?
(in response to question c.)

Old-timers will recall with a touch of nostalgia the lament of the early 1960s,

"Where have all the flowers gone?" In this first decade of the 21st century, cosmologists bewail a similar refrain, "Where has all the anti-matter gone?"

Despite valiant and heroic attempts to detect its presence, our Universe seems to be devoid of anti-matter. This presents a major problem for the Standard Model of the Big Bang. Some authors even suggest that the theory is on shaky ground until we can come up with a consistent answer to this question (Ref. 1.10).

Most scientists agree that according to the Standard Model of the Big Bang, an infinitesimal singularity of infinite energy density exploded, converting part of this energy to equal amounts of matter and antimatter within an expanding bubble of space-time. As the Universe expanded, the energy density decreased with a corresponding drop in temperature, from trillions of degrees to the present 2.7 degrees Kelvin of the CMBR (Fig. 1.3). This log-log plot

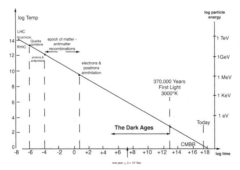

Fig I.3 – The standard model for the temporal evolution of the Universe after the Big Bang

of temperature as a function of elapsed time graphically shows the sequence of events in the evolution of the Universe following the Big Bang with numbers in powers of 10. As the temperature fell, the equilibrium between production and recombination of matter and antimatter shifted toward recombination.

Temperature is another way of expressing the kinetic energy of the particles. For example, the decrease in temperature in the time interval from 10^{-5} seconds to 5 seconds is known as the epoch of matter-antimatter recombination. The decrease from 100 billion degrees to 500 million degrees corresponds to a reduction in energy from 2 GeV to 1 MeV, corresponding to the thresholds for pair production for protons and electrons respectively (see introductory comments of Appendix 1.4 for explanation of the eV as a unit of energy). Thus, the proton-antiproton annihilation occurs at the beginning of this epoch, whereas the recombination of electrons and positrons is completed at the end, corresponding to their respective energy gaps. The big science mystery is simply stated:

1. At the Big Bang, equal amounts of matter and antimatter were created.
2. All forms of matter and antimatter were mutally annihilated within 5 seconds of the Big Bang.
3. So, where did the excess matter come from to start the process of nucleosynthesis (hydrogen and helium nucleii) over the first three minutes and then go on to aggregate, forming the first stars and galaxies?

Several theories have been proposed to explain this asymmetry, such as a slight difference in decay rates between matter and antimatter resulting from a process known as CP violation. This would lead to a slight excess of matter over antimatter to one part in 10 billion before their mutual annihilation. In this author's opinion, this argument seems to resemble the original fudge factor of Einstein's cosmological constant, which he proposed to maintain his model of the steady state universe.

Recent experiments in high energy colliders at Los Alamos and Fermilab indicate that there may be some difference in these decay rates, but not enough to account for the imbalance required by the standard model. These experiments on the dynamics of neutrinos were designed to provide a clear answer to the suggested imbalance between matter and antimatter. One of the researchers reporting on these results remarked, "I was sort of expecting a clear excess or no excess – in a sense we got both." Like the numerous satellite attempts to detect gamma ray evidence for the presence of anti-matter, the results are inconclusive or even negative (Ref. 1.11).

Other suggestions have been put forth that matter and antimatter were physically separated before and during the rcombination epochs by some unspecified and mysterious mechanism. As a consequence, some believe that somewhere beyond the horizon lie undetectable gigantic galactic superclusters of antimatter. As one author writes, "Is it possible to think of a mechanism that separated matter from antimatter during the time the Universe was very hot and dense? Apparently not" (Ref. 1.12).

So, how does all this help us resolve the contradiction posed in question c. on the different chronology in our interpretation of the Day-Age model of the Bible and present scientific theory? Unfortunately, it actually serves to deepen the mystery. Consider once again the description of the Second Day in the Book of Genesis 1.6:

"God said let there be a *Rakia* in the midst of the waters." We have previously conjectured that this separation corresponds to the band gap between the unobservable negative energy states and the positive energy states of our world, and consequently serves as the divide between matter and anti-matter. In view of the suggestions by some that a permanent physical segregation between matter and antimatter took place in the early seconds after the Big Bang, it is tempting to speculate that the *Rakia* is such a physical barrier between our Universe and a parallel anti-matter universe. Most amazing is some of the language used by our commentators as they try to shed some light on the pithy description of the Second Day in the Bible (Ref. 1.13):

A. Rashbam – "the Rakia segregated the waters leaving half above and half below," clearly evocative of a process in which equal amounts of matter and antimatter would have been separated into their respective universes.
B. B'chor Shor – "'let it be a divider' denotes that the division is intended to be permanent."
C. Malbim – "the barrier between the waters was not only a division in space but also in kind," i.e. between matter and anti-matter.
D. Midrash – as noted previously, it states unequivocally that, "…were it not for the *Rakia*, the earth would be engulfed by the waters above and below," certainly evocative of a cataclysmic collision between matter and anti-matter.

Appendix 1.5 gives additional background information on these sources.

Apparently, the Torah and its interpreters seem to provide an answer to the question, "Where has all the anti-matter gone?" by suggesting that an impenetrable barrier was established in the early seconds of the Big Bang, forever separating our material Universe from its anti-matter twin. However, we are not told how this may have occurred within the framework of causality and the laws of physics as presently understood and accepted.

The chronological problem posed by question c. is explicitly illustrated in Fig. 1.3, which plots the drop in temperature as a function of time after the Big Bang on a log-log scale. Several epochs are identified as the Universe expanded according to the Standard Model (Ref. 1.14).

According to our interpretation, the barrier between matter and anti-matter is described as occuring during the Second Day of Genesis – some time after the appearance of the First Light on the First Day, which we have identified as the CMBR in Sections

1.B.4 and 1.B.5 of this Chapter. However, according to the Standard Model, this First Light appears some 370,000 years later. If we wish to retain the Day-Age model, it will be necessary to modify one or more of our speculations as follows:

1. At the end of Day One of Creation, the First Light was not the escaping photons from the 3000 degree plasma of electrons and hydrogen ions as described in 1.B.4. In order to fit the chronology, it would need to be something that was produced and then escaped from the primordial soup during the first seconds after the Big Bang; could it be that the present diffuse soft x-ray background may be a downshifted remnant of the proton-antiproton annihilations occurring during the first second, when some high-energy gamma rays managed to escape? The energy spectrum of this diffuse soft x-ray background extends from 0. to 2 keV, downshifted by a factor of one million from the 2 BeV gamma rays, which may have been emitted at the onset of the recombination epoch, and by a factor of 20,000 from the one MeV photons during the final electron-positron recombinations.

   > It has been suggested by my colleague Pinchas Rosenfelder that there were at least two cycles of darkness and light until the formation of the Earth on Day 3, e.g. "...it was evening and morning, Day One." (Cosmic x-Ray Background Radiation – CXBR in analogy to CMBR)

2. Our interpretation of the raqiya as the band gap and barrier between matter and antimatter is not correct.
3. The Standard Model is not correct and needs to be modified; was there indeed a First Light in the first five seconds?
4. Many commentators posit that there is no past or future in the Torah, i.e., it is not written in chronological order. However, several authorities say that this rule does not apply to the Six Days of Creation. In fact, in a 17th century work, Me'am Loez states that following the separation of light from darkness, the Torah proceeds to a chronological description

of the rest of Creation – certainly the basis of the Day-Age model (see information on Me'am Loez in section D.2).

It is most appropriate to summarize with a quote from Ramban, "...the events of the Second Day deal with the most innermost mysteries of Creation and I will not comment further" (Ref. 1.2).

Appendix 1.4 – The Wave Equations of Schrödinger and Dirac: Matter & Anti-Matter

The history of the development of Wave Mechanics and its place in Modern Physics is both exciting and informative. In the early years of the twentieth century, Max Planck was able to explain the Black Body Radiation Spectrum by invoking a "quantum of action" (h). He postulated that the oscillators (atoms) in the black body could only emit light by making discrete quantum jumps from one state to the next. This was a revolutionary break from the classical physics of Newton and Maxwell.

The energy (E) of each transition was given as: $E = h\nu$, where ν is the frequency of the emitted light. The value of h (Planck's Constant) is 6.634×10^{-34} Joule-seconds (in MKS units). For a single oscillator transition emitting blue light photons (of frequency 1015 cycles per second) the energy is 6.6×10^{-19} Joules, an infinitesimal amount in our macroscopic world. One watt of power is one Joule per second. An electron-volt (eV) is the energy given to an electron when accelerated by an applied voltage of one volt. Since the charge on an electron is 1.6×10^{-19} Coulombs, the value of h in electron-volt .sec is 4.2×10^{-15}. The energy of these blue light photons could therefore also be expressed as 4.2 eV. Mankind's view of the physical world is intermediate to the cosmic scale of our Universe and the submicroscopic scale of the atom world. Therefore, Newtonian Mechanics is an approximation which works well for us, but which breaks down when applied to either galaxies or the interior of atoms. Einstein's Theories of Relativity were initially used to explain and predict phenomena on a cosmic scale, whereas quantum (or wave) mechanics were needed for the submicroscopic world.

In 1905, Albert Einstein used Planck's concept to solve the mystery of the photoelectric effect (an ejection of electrons from the surface of a solid when exposed to light).

No matter how intense the light, no electrons would be emitted unless the light frequency exceeded a threshold dependant on the target material. In his paper (published in the same year as his Special Theory of Relativity), Einstein introduced the concept of duality wherein light propagates as a wave, but also acts as a corpuscular stream of "photons" in its interaction with matter (Ref. 1.15).

Approximately twenty years later, Louis de Broglie applied duality to matter in order to make sense out of the restricted electron orbits in the Bohr-Rutherford atom. Only those wavelengths which gave constructive interference in the orbit were permitted (Fig. 1.4).

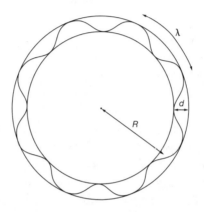

Fig I.4 – A de Broglie wave representing a particle. For the wave to "join up with itself" smoothly, we need $n\lambda = 2\pi R$, where n is an integer.

(d is the "thickness" of an atomic orbit and is considered to be infinitesmal)

Schrödinger two years later published the Wave Equation which bears his name, introducing the concept of the wave function Ψ, which describes the probability of finding a particle at coordinates (x, y, z) at time t. The solutions to this Schrödinger Wave Equation automatically give the energy levels and orbits of the Bohr-Rutherford atom.

In 1928, Paul A.M. Dirac developed a relativistic wave equation whose solutions led to some astonishing conclusions. The energy eigenvalues (states) for electrons and other particles would have two possible values given by

$$E = \pm c (p^2 + m^2 c^2)^{1/2}$$

where mo is the rest mass, c is the velocity of light and p is the momentum.

For a free particle at rest, p=0. and there are two possible values:

$$E = \pm m_0 c^2$$

The solutions of the Dirac Equation allowed a particle to have negative kinetic energy. Dirac proposed that we regard these negative energy states as all filled from the rest mass energy to minus infinity. The Pauli Exclusion Principle dictates that each of these states is occupied by only one electron. In the language of quantum mechanics, no two particles can have identical quantum numbers, which is equivalent to saying in plain language, "no two things can occupy the same space at the same time." As a consequence, the normal state of empty space consists of an infinite density of negative energy electrons.

According to Dirac, all of these negative energy states are occupied, from energy of $-m_0c^2$ all the way down to minus infinity. Quantum transitions are allowed between negative and positive energy states as long as the input entity (like a gamma ray) has enough energy to promote electrons across the forbidden gap (Fig. 1.5).

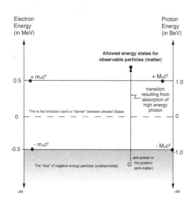

Note 1: $M° = 1840\ m°$ Note 2: The center of the forbidden band may be analogous to Rashi's בתוך המים (in the midst of the waters)

Fig I.5 – Allowed solutions (energy states) of the Dirac Wave Equation

The "sea" of occupied negative energy states is separated from the observable world of positive energy states by a forbidden band gap of twice the rest mass energy. A physical manifestation of their virtual existence only occurs when one of these electrons is promoted across the forbidden gap of $2\ m_0c^2$, leaving a hole in the universe of negative energy electrons by a process called "pair production." The rest mass energy of the electron is 0.51 MeV (million electron volts), so the pair production requires an input energy of 1 MeV. The rest mass of the proton is approximately 2000 times that of the electron, so the band gap for this particle is 2 BeV (two billion electron volts). Modern usage writes this as 2 GeV.

This hole in the sea of the filled states manifests itself as a positively charged particle (anti-electron or positron) that has the same mass as the missing electron.

In 1932 David Anderson observed the creation of electron-positron pairs in his cloud chamber experiments in his "lab" at the top of the aeronautics building at the California Institute of Technology in Pasadena. This location had the necessary electrical power to energize the big electromagnets employed by Anderson.

A high-energy gamma ray from outer space entered the cloud chamber and "promoted" one of the negative energy electrons across the energy gap. His photographs showed two diverging tracks (due to this applied external magnetic field), one track being the promoted electron and the other track the hole in the sea of negative energy states – a positive particle called a positron. The rest mass of the electron has an energy equal to 0.51 MeV, so that the γ-ray that produced the pair had at least 1.02 MeV energy. For his discovery, Anderson was awarded the Nobel Prize in Physics, at the age of 31.

Since that time, positrons have been routinely observed in experiments carried out with high-energy accelerators and more recently anti-protons and other anti-matter particles have been produced. An anti-matter hydrogen atom would then consist of a negatively charged anti-proton at the nucleus and a positively charged electron (positron) in orbit around it. In 1995 such anti-matter hydrogen atoms were produced in the laboratory, lasting about 40 nanoseconds before annihilation. When an anti-particle encounters its opposite entity, both are annihilated. The observable normal electron has returned to the hole in the invisible sea and the energy is emitted as a high-energy photon.

Hidden Light: Science Secrets of the Bible

Appendix 1.5 – Talmudic and Rabbinic Exegetes
Ibn Ezra: R. Avraham b. Meir Ibn Ezra (1093–1167)
Hebrew poet and biblical commentator
B'chor Shor: R. Joseph B'chor Shor (second half of 12th century)
French Tosafist and exegete
Rashbam: R. Samuel b. Meier, grandson of Rashi (1085–1174)
Biblical exegete
Malbim: R. Meier Loeb b. Yehiel Michael (1809–1879)
Hebraist
Ba'al Haturim: R, Jacob b. Asher (first half of 14th century)
German author on Jewish law and biblical commentator
Pirkei d. Rabbi Eliezer: An aggadic midrash, in its present form originating from the 8th or 9th centuries

D. RESOLUTION OF THE CONTRADICTION BETWEEN THE CHRONOLOGY IN GENESIS AND COSMOGONY – FOR DAYS THREE AND FOUR (SUN, MOON, STARS APPEAR ON DAY 4, AFTER FORMATION OF EARTH ON DAY 3)

1) The Nature of the Primordial Atmosphere

Planet Earth was created on the Third Day, together with the seas and all vegetation. On the Fourth Day, God says: "Let there be luminaries in the expanse of the heaven (ברקיע השמים) to provide a division between the day and the night and they [the luminaries] shall be for signs and for seasons and for days and years."

Rashi here quotes the Gemara in *Chagiga* 12a, which states that the sun, moon and stars were part of the creation of Day One and that it was only on the Fourth Day that they "appeared".

Others claim that the stars, including our sun, were formed at the end of the Second Day. Table 1 shows stellar evolution to have occurred at the transition between the Second and Third Days (about 200 million years after the Big Bang). This is an arbitrary assumption to match the time scales. Commentators have also noted that the word אותות (signs) signifies the use of the sun, moon and stars as guides to navigation.

Based on a suggestion by E.G. Freudenstein, Prof. Joseph Bodenheimer of the Jerusalem College of Technology has proposed that we consider the condition of the primordial atmosphere which enveloped the earth on the Third Day (Ref. 1.16).

The newly formed earth's atmosphere was an opaque mix of hydrogen, water vapor, CO_2 and gases like methane. Both scattering and absorption of the sun's light was the order of the "day".

In Genesis 11.4, the creation is recapitulated as follows:

"These are the *Toledoth* (אלה תולדות) of the heavens and the earth when they were created...," usually translated as:

"These are the generations of the heavens and the earth..." But many other meanings have been proposed for *Toledoth*, including offspring, products, and inherent potential developments.

Two sentences later, Genesis 11.6 states that "...a mist rose from the earth and watered the whole surface of the ground." The account of creation is written as if a hypothetical human observer was present at these events. For a man standing on the surface of our planet and gazing skyward on the Third Day, the already-created heavenly host would not be visible to him. The mist, consisting mostly of H_2O in the form of water vapor, would probably have been thicker than the densest fogs of today, with a scattering attenuation of more than 400 dB/km, corresponding to a visibility of less than 50 meters. On the other hand, the scattered sunlight would be able to reach the plants and initiate the photosynthesis process wherein water vapor and carbon dioxide are combined to produce starches and sugars in the vegetation:

$5\ CO_2 + 5\ H_2O = C_5H_{10}O_5 + 5\ O_2$.

The chlorophyll in plants acts as a catalyst to produce starch and higher order carbohydrates under the stimulus of sunlight.

During a period of time extending over the Third Day, the dense water vapor molecules and the carbon dioxide were used up and oxygen was released into the atmosphere. The air cleared and, presto, the luminaries (sun, moon and stars) appeared on the Fourth Day (to our hypothetical observer).

It is also conceivable that a small fraction of free atmospheric oxygen was produced in the photo-dissociation of water by the ultraviolet portion of the solar spectrum.

Only after the plants produced enough oxygen (O_2) by photosynthesis did the protective screening action of the oxygen isomer, ozone (O_3), make life possible on our planet, as well as provide a transparent atmosphere for viewing the heavenly host.

2) Problems with Rashi's Commentary on Genesis 11:5–8

Rashi's comments on these sentences are at variance with the model of atmospheric cleansing as presented above. He writes, "But as regards the Third Day of Creation about which it is written 'The Earth brought forth sprouts, herb yielding seed after its kind and tree yielding fruit…,' this does not signify that they came forth above the ground but that they remained at the opening of the ground (i.e., just below the surface) until the Sixth Day." If so, it is hard to see how photosynthesis could take place. Rashi propounds this hypothesis in order to literally and Figuratively prepare the ground for the creation of Adam on the Sixth Day.

It could be that Rashi was somehow anticipating some of the latest discoveries on the sources of oxygen in photosynthesis. The previous discussion implies that oxygen originates from plant leaves as they produce starches and sugars by using up the carbon dioxide and water vapor in the atmosphere. Using radioactive tracer techniques, researchers have established that the roots of the plant play a significant role in releasing oxygen by decomposition of the water. The carbon dioxide and hydrogen are then used by the plant to manufacture its food with the action of sunlight, i.e. the oxygen does not directly result from the consumption of carbon dioxide.

A number of commentaries differ with Rashi on the creation of the sun, moon and stars. Some claim that these luminaries were formed during the Second Day, while *Me-am Loez*[5] states that it was on the Fourth Day, as follows:

"God created the earth and all its properties on the Third Day and only afterwards, on the Fourth Day, did he create the sun (to dispel any notion that creation of Earth was a natural result of the sun's heat)." These words of *Me-am Loez* were written almost 300 years ago, well before the modern scientific era with its theories of cosmogony (origins of the solar system). They serve as an example of the value of scientific knowledge to avoid making mistakes in Biblical exegesis. It is quite likely that 300 years into the future, exegetes will view our current attempts with bemusement.

The use of the term ברקיע השמים (translated as the expanse of heaven) to denote the abode of the luminaries has drawn the following comment from the Midrash: "…these concepts are beyond Man's grasp. It is an exceedingly difficult matter and no mortal can

---

5. *Me-am Loez* is an 18th century work written in Ladino by Jacob Cali (1730) and his successors for people (Sephardim) in Turkey and the Balkans who had little or no access to the sources (Talmud, Zohar, etc.). Later it was partially translated into Arabic for those living in North Africa. The title of the work comes from Psalm 114 referring to the Exodus from Egypt (…*from a people of a strange language*).

fathom it." It is hoped that with the insight provided by the science of the 21st century, combined with intensive study of Biblical texts, we may be granted a better understanding of these mysteries.

E. THE AGE OF THE UNIVERSE

1) Does the "Age" Depend on Location?

Before we enter into a discussion on the age question, the effect of location on the answer needs to be considered.

Special Relativity teaches that objects in relative motion will have their clocks running at different rates, and General Relativity tells us that the flow of time also depends on the gravitational field of the observer. Consequently, one might easily surmise that any inhabitants of distant galaxies (which are rushing away from us at high speed) will get different answers to the age question than people on Earth. There are two observations which lead to the conclusion that there is in fact a universal measure of time, i.e. clocks everywhere in our cosmos are ticking away at the same rate:

1. The time dilation caused by relative motion is the result of such motion *through* space. However, it is space itself which is expanding and the consequent increasing separation of the galaxies will not result in time differences between them – the galaxies are like raisins embedded in a cake in the oven which is rising, or, as some prefer, the galaxies are like rigid buttons pasted on an expanding balloon.
2. The thermal uniformity of the CMBR (Cosmic Microwave Background Radiation) as determined by WMAP (Wilkinson Microwave Anisotropy Probe, launched in 2001) suggests that physical conditions at the macro level were the same everywhere in the evolution of the Universe.

We therefore conclude that the question of the Age of the Universe would have the same answer, regardless of location.

2) A 13th Century Calculation

Rabbi Isaac of Akko (Acre) was a follower of the Ramban (Nachmanides), and was one of the foremost Kabbalists of his time (1250–1350). Rabbi Isaac carried out a remarkable calculation some seven hundred years ago, based on the *Sefer ha-Temunah* ("The Picture Book"), an ancient work which speaks about "Sabbatical Cycles". According to this text, there were other "worlds" before the creation of Adam. In the language of a related Midrashic commentary, "God created universes which were subsequently destroyed." Note Sanhedrin 97a, a tractate of the Talmud, where it states, "...the world will exist for 6000 years and in the 7000th year it will be destroyed." The *Sefer ha-Temunah* relates that there are seven Sabbatical cycles comprising 7000 years each, wherein the "world exists for 6000 years and in the 7000th year it is destroyed." The year 6000 of the Hebrew calendar will be the end of the present cycle (about 230 years hence). This concept of Sabbatical Cycles was derived from Leviticus XXV:8–10, which instructs the Children of Israel on the laws of *Shmita* (allowing the land to lie fallow every seventh year) and Jubilee, commencing 10 days after the 49th year on Yom Kippur of the 50th year (see Appendix I.6). In modern-day Israel, Shmita years (such as the year 5768 (2007–2008)) provoke much discussion and even disputation on proper observance of this Biblical injunction, taking into consideration the economic impact on local farmers and produce providers.

Rabbi Isaac goes on to make the analogy that just as there are seven times seven years preceding a Jubilee, there are seven Sabbatical cycles in the evolution of the Universe. Each Sabbatical cycle is analogous to the seven earth years of **Shmita**.

It should be emphasized that this concept of Sabbatical cycles was vigorously disputed by a number of rabbinical authorities, especially Rabbi Moses Cordovero (also known as the *RaMaC*) and Rabbi Isaac Luria (the *Ari*), as well as Saadia Gaon. Aryeh Kaplan (1934–1983) points out that since this is not a matter of law or faith, there can be no binding opinion from either side of the controversy (Ref. I.7) Rabbi Kaplan had a background in both physics and Judaism. He studied at the Mir Yeshiva in Jerusalem and was ordained by leading rabbinic authorities in Israel. During his studies for a Master's degree in physics, he was described by a "Who's Who" in science as the most promising young physicist in America.

There are also various opinions on the present cycle amongst those rabbinical authorities who do adhere to the concept of Sabbatical cycles. One opinion says we are in the seventh cycle, which means the Universe would have been 42,000 years old at the creation of Adam. Rabbi Isaac proceeds to combine this number with a well known interpretation of line 4 in Psalm XC to calculate the age of the Universe.

א. תפלה למשה איש־האלהים אדני מעון אתה היית לנו בדר ודר;
ב. בטרם הרים ילדו ותחולל ארץ ותבל ומעולם עד עולם אתה אל;
ג. תשב אנוש עד דכא ותאמר שובו בני אדם;
ד. כי אלף שנים בעיניך כיום אתמול כי יעבר ואשמורה בלילה.

1. *A prayer by Moses, the man of God: My Lord, a dwelling place have You been for us in all generations.*
2. *Before the mountains were born and You had not yet convulsed the earth and the inhabited land, and from before the world to the end of the world You are God.*
3. *You reduce man to contrition and You say, "Repent, O sons of man."*
4. *For a thousand years in Your eyes are but a bygone yesterday, and like a watch in the night.*

Since the six previous sabbatical cycles existed before the present period, their chronology should be measured in divine years. Line 4 of the psalm states that 1000 years on Earth is equivalent to one day in the sight of God. Therefore a divine year is 365,250 years as experienced by us on Earth. By multiplying the number of divine years prior to the present cycle (42,000) by the number of Earth years per divine year (365,250), Rabbi Isaac achieved the incredible result of the age of the Universe as 15.3 billion years. Incredible, since this number is amazingly close to the values accepted by most cosmologists up to the year 1995, and which appears in most books published before that date.

After the early results of the Hubble Space Telescope were analyzed in 1995, the age was reduced somewhat, placing it in the range of eight to twelve billion years, and it appeared that Rabbi Isaac's calculation would be relegated to the dustbin as an historical curiosity.

This downward revision was based on the measurement of the distance to galaxy M100, about 56 million light years away from Earth. It caused great puzzlement to astronomers and cosmologists, since there are stars in our own galaxy (the Milky Way) that are believed to be older than 12 billion years. The discrepancy is apparently due to conflicting results from two different approaches to the research: stellar evolution theory based on brightness measurements (especially the "hot" stars in globular clusters near the end of their life cycle) vs. the distance scale measurements.

The latest results from WMAP (Wilkinson Microwave Anisotropy Probe) corrected the Hubble Telescope measurements, producing a value of 13.7 ± 0.2 billion years. Rabbi Isaac's calculation from some 700 years ago is thus an amazing achievement, deviating less than 12% from the most recent scientific measurements.

3) The Age of the Universe in Christian Thought

Early Christian scholars also had varying opinions on this subject. For example, Justin Martyr and Irenaeus in the second century (1000 years before Rabbi Isaac) gave their own interpretations of Psalm 90, saying that each "day" of Genesis corresponded to a thousand years. Origen talked of a spiritual meaning to the first three "days", and only on the fourth day, with the appearance of the luminaries, did Time (as we count it) begin. In his work "The Literal Meaning of Genesis", Augustine of Hippo (354–430) wrote: *"But at least we know that it* [the Genesis creation day] *is different from the ordinary day with which we are familiar."* Augustine was born in Algeria to Berber parents. His father was pagan and his mother was Catholic. As a youth, he pursued a hedonistic life style and consequently composed one of his more famous prayers "Grant me chastity and continence, but not yet." He insisted (despite great opposition) to read and write in Hebrew, as well as Latin, and wrote extensive commentaries on scripture. One of his most famous works (over 1200 pages long), "The City of God against the Pagans," contains detailed analyses of all the Psalms.

Bishop Ambrose (4th century) postulated that the six days of creation corresponded to a 144-hour period and this view predominated, culminating in the calculations of John Lightfoot, Vice Chancellor of Cambridge, and James Usher (1642 to 1650 – about 30 years after the issuance of the King James Bible), that all creation took place during the week of October 18–24, 4004 BCE. This uncompromising position has been a major factor in the ongoing conflict between the religious and secular world views.

4) An Alternative Calculation

The approach described in this section was discussed by the author more than a decade ago (with his sons and grandchildren), so it should not be viewed as working backwards from the WMAP results to get a better Biblical value.

Consider once again the full sentence in Psalm XC:4:

"כי אלף שנים בעיניך כיום אתמול כי יעבר **ואשמורה בלילה**"
"For a thousand years in Your eyes are like a bygone yesterday and like a watch in the night."

The phrase "like a watch in the night" has been apparently neglected in all previous discussion on the subject, with the possible exception of Augustine (see his exposition on Psalm 90 in *The City of God*). It will be shown that a careful consideration of this phrase can lead to a determination of the number of earth years per divine year, without resorting to Sabbatical Cycles to calculate the age of the Universe.

The original calculation, as described in the previous section, gives a result of 365,250 earth years per divine year, and is based on one divine day for every thousand years. However, the phrase "like a watch in the night" provides a multiplier – instead of one divine day per thousand years, it could be read as one divine watch in the night per thousand earth years. There is some argument amongst the commentators whether a watch in the night extends over four hours or six hours, with four hours as the preferred number. As one who stood guard duty in the Scouts and later in the U.S. Navy, I also lean towards the four-hour number (which means that there would be six watches per 24-hour day). The new result for conversion from divine to earth years is then $6 \times 365{,}250 = 2.1915$ million earth years per divine year. If there were 6000 divine years (corresponding to the six days of Genesis) before the creation of Adam and God abstained from his creative activity at this point, then the age of the Universe would be: $6000 \times 2.1915 \times 10^6$ years, or 13.149 billion years. This value differs from WMAP by 4%.

This alternative interpretation of Line 4 in Psalm 90 provides the following:

A. It obviates the need to employ the controversial Sabbatical cycles in estimating the age of the Universe.
B. There seems to be a logical contradiction in the use of the 42,000 divine years when computing the age of *this Universe*. During these 42,000 divine years, six prior "worlds" were created and destroyed.
C. It produces a value closer to the latest scientific data.

These same Sabbatical cycles may also contain a clue to the fate of the Universe, as described in the following section F.

F. THE FATE OF THE UNIVERSE: EXPANSION FOREVER OR THE BIG CRUNCH?

1) Philosophers and Poets

> "*And so some day,*
> *The mighty ramparts of the mighty universe*
> *Ringed sound with hostile force,*

> Will yield and face decay and come crashing to ruin."
> — Lucretius, De Rerum Natura

> "Some say the World will end in Fire, others say in Ice/ From what I've tasted of desire/I hold with those who favor fire."
> — Robert Frost, "Fire and Ice"

> "This is the way the World ends,
> Not with a bang, but a whimper."
> — T.S. Eliot, "The Hollow Men"

Section E focused on the past and present history of the Universe. As expressed in the above quotations, most view its future with great pessimism. In his book "The Last Three Minutes", Paul Davies quotes Hermann von Helmholtz, who wrote in 1856 that the Universe is dying "a heat death" as a consequence of the Second Law of Thermodynamics (the inevitable and irreversible nature of entropy, which only increases in closed systems) (Ref. 1.18). This depressing prediction has exerted a profound influence on many scientists and philosophers. Davies quotes Bertrand Russell, who wrote in a book entitled "Why I Am Not A Christian," "*All the labours of the ages, all the noonday brightness of human genius, are destined to extinction in the vast death of the solar system.*" Actually, the death of the solar system will occur long before the heat death of the Universe. The Sun is currently about half way to extinction; according to the standard model, it was formed about five billion years ago and will continue to provide Earth with life supporting radiation for another five billion years before becoming a red giant.

We noted in section B that unlike most other galaxies, Andromeda is approaching Earth rather than receding from it. Astronomers estimate that there will be an intergalactic collision between the Milky Way and Andromeda in about 3.5 billion years, two billion years before the sun engulfs the solar system. However, as dramatic as this may sound, most believe that this will be a peaceful intermingling due to the great interstellar voids.

In Section B.5, Edgar Allan Poe's prose poem "Eureka: An

Essay on the Material and Spiritual Universe" was cited as the first explanation for Olbers' Paradox, based on Poe's concept of an expanding Universe (from 1848 – exactly 100 years prior to the papers by Gamow and his group on the Big Bang). At the top of the manuscript, Poe wrote his dedication: "With very profound respect this work is dedicated to Alexander von Humboldt," a reference to the first recognized environmentalist. Von Humboldt was a masterful naturalist and explorer, whose name marks the famous current off the western coast of South America. Poe continues a few sentences later: "My general proposition then is this – In the Original Unity of the First Thing lies the Secondary Cause of all Things *with* the Germ of their Inevitable Annihilation."

In the following sections, the latest scientific models for the ultimate fate of the Universe will be described. It will be shown that the gloomy Helmholtz prediction is only one of several possibilities. The chapter concludes on an optimistic note with a discussion of the Biblical sources in light of the current scientific data.

2) The Friedmann Solutions to the Einstein Equations

As noted in Section B.2 (Steady State or Big Bang), Alexandr Friedmann published his version of Einstein's Equations as applied to the cosmos in 1922 (about seven years before Hubble's results saw the light of day). Several important parameters of the Universe which appear in these differential equations determine the nature of his solutions, resulting in three distinct models for the past, present and future of the Universe. These parameters are:

* R is the "scale factor," or radius (size) of the Universe, which varies with time, t [R(t)]
* P is the mass density
* k is the shape or curvature of the space-time

Although these equations are not easy to solve, Friedmann came up with three special solutions, each corresponding to a different curvature of space-time; that is the value of k. The three solutions are graphically illustrated in Fig. 1.6, where the "size" of the Uni-

Hidden Light: Science Secrets of the Bible

verse, R(t), is plotted as a function of time for k = -1, 0, +1. The value of k depends on P, the amount of mass density in the Universe, relative to a critical "mass density".

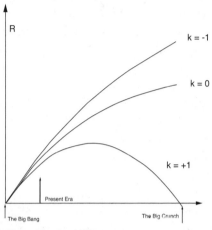

Fig. I.6 – Friedmann's Solutions of Einstein Equation

The situation is quite analogous to the concept of escape velocity from Planet Earth, which has become quite familiar to all of us in the Space Age. The value of approx. 11 km. per sec. to escape from the gravitational pull of the earth's mass corresponds to the k = -1 curve. In other words, the energy of the Big Bang in this case was more than enough to overcome the gravitational "pull" of the mass density so that the expansion goes on "forever". The mathematicians call this a case of an open hyperbolic universe (curved like a saddle).

On the other hand, for k = +1 there is sufficient mass density (maybe even dark matter) to overcome the initial acceleration of the Big Bang and the Universe starts to collapse after achieving its peak expansion, much as a rocket falls back to earth from its peak trajectory because it didn't achieve escape velocity. This Friedmann model corresponds to a spherical closed space-time manifold. However, at the Big Crunch (about 25–35 billion years from now), some theorists propound the idea that it starts all over again, leading to the concept of an oscillating universe.

The intermediate case, k = 0, corresponds to a mass density equal to the "critical density".

The expansion continues, but at a slower rate than the k = -1 model. This model corresponds to a "flat universe" in space-time.

In the two models of open and flat universes, the "heat death" of Hermann von Helmholtz is a likely end.

For many years, there was a great amount of controversy over which model resembled reality. The big unknown was the mass density – how much matter is there? Most of the estimates varied, but were never more than ± 10% of the critical mass density, P_c. Recent observational data from WMAP suggest that P is equal to P_c with an error less than 3%. However, this does not necessarily signify that we inhabit a flat universe. It only means that the space scanned by WMAP is flat. Beyond the "horizon," space can still start to curve (just as the earth appears flat to us, but starts to curve beyond our horizon).

3) Biblical Sources

Although there are no explicit references to the eventual fate of the Universe in the Bible, certain recurring phrases may provide some insight into this question. There is a hint in the last sentence of the second part of the *Shema*, which concludes with the following, "In order that your days may be prolonged and the days of your children upon the land which Hashem swore to your fathers to give it to them as long as the sky remains above the Earth." Unlike the predominant beliefs of the ancient world, which held that the Universe was static and eternal, this passage from the Book of Deuteronomy foresees a time when there is no more Earth. All of the phrases contain the word "*OLAM*" (עולם), previously encountered in Sections B.4 and 5, where the Cosmic Microwave Background Radiation (CMBR) was identified with the Light of Genesis. There is an evocative similarity with the term *YLEM* bestowed on the primordial substance of the Big Bang by Gamow, Alpher and Herman. Simply put, *OLAM* is considered by this author as the entire Universe in space-time. In addition to its appearance throughout the Bible, three versions of the word are found in the most frequently recited prayer of the Jewish liturgy, the "Kaddish".

The Kaddish prayer in its several forms is recited only in the presence of a *minyan* (ten or more men), several times at each of the three daily services and also on Sabbath and Festivals. It is written and recited in Aramaic, but its origin has been obscured by the mists of antiquity. Some say that it was introduced by the Great Assembly (Sanhedrin) in the early days of the Second Temple (around 4th century BCE). Other authorities believe it predates the destruction of the First Temple since there is no mention of Jerusalem.

After the reader chants the opening lines of the prayer, in testimony and praise to the greatness and holiness of God's name, the entire congregation responds in unison:

"יהי שמי רבא מבורך לעלם ולעלמי עלמיא," usually translated as "May His Great Name be praised forever and ever." As I interpret the phrase, it means something much more. Before proceeding to explore this deeper meaning and its possible relevance to our discussion, some related Biblical sources will be examined.

The first part (May His Great Name be praised) is mentioned in Deuteronomy XXXII:3. On the last day of his life, Moses asks the assembled Children of Israel: "Lend me your ears." He then declares, "I will call upon the Name of the Lord; ascribe ye greatness unto our God." A similar expression is found in the Book of Job (Job I. 20–21). He has just learned of the great disasters which have befallen his family and the total loss of his fortunes. In spite of all this tragedy, he utters the lines: "The Lord giveth and the Lord taketh away; may the Name of God be blessed."

A number of Psalms, probably composed before the Kaddish prayer, contain references to *OLAM* and variations thereof. Psalm XC:2 quoted in Section E.2 contains the phrase "ומעולם עד עולם אתה" – usually translated as "…from this world to the world to come You are God." Others translate it as "forever and ever" and some say from "everlasting to everlasting." The same expression is found in I Chronicles XVI:36. Psalm CXIII:2 is translated as "May the Name of God be praised from now and forever," very similar to the Kaddish.

The closest source to the full expression is in the Book of Daniel. Daniel was among the captives brought to Babylon by King

Nebuchadnezzar before the final destruction of Jerusalem. The King had a dream which he forgot (or pretended to have forgotten). In a fit of rage at the inability of his "wise men" to relate and explain the dream, he issued a royal decree to destroy them all, including Daniel and his friends. Daniel prayed fervently to God and the secret of the dream was revealed to him in a vision of the night.

In gratitude, he blesses the God of Heaven, declaiming in Aramaic,

"שמי די אלהא מברך מן עלמא ועד עלמא" – variously translated as "Blessed be the Name of God from everlasting to everlasting" or "from the earthly to the heavenly spheres."

A beautiful Aramaic poem called "Ya Ribon" is sung by many during the festive Sabbath meal. It was composed by Rabbi Israel Ben Moshe of Najara, a student of "the Ari" in Safed, and later to become the Rabbi of Gaza. The first stanza begins with "Ya Ribon Olam v'Olmaya...," translated here as "O, Master of this Universe and all Universes to be." This is not a translation one usually finds in the bilingual song books, where it appears as "Lord of all worlds" or "Master of this world and all worlds." Similarly, the three key words in the Kaddish prayer – L'Olam, U'Lolmay, Olmaya may now be interpreted as referring to multiple or even parallel Universes, which leads us to a renewed consideration of the Friedmann models.

4) Science and Eschatology (the End of Days)

We'll begin with Robert Frost's prediction, "...the world will end in fire," which corresponds to the case of $k=+1$ (a closed spheroidal universe). In this model, the Universe expands to a maximum size and then begins to contract. For an astronomer of that time far into the future, the first indication of trouble ahead will be a change from the red shift of nearby galaxies to a blue-shifted spectrum. This will be happening long after our sun has become a red giant, engulfed the earth and solar system and then burned out (about five billion years from now). Our fictitious astronomer might be the descendant of those refugees from Planet Earth who made it to another part of our galaxy's inhabitable zone. A report in *Scientific American* on October 4, 2007, details the discovery of an "Earth-Like Planet in the

Making," circling a star 424 light-years from us. This represents the first case of a potentially inhabitable planet in our galaxy, following observation of more than 200 planets, all of whom appear to be gas giants like Jupiter and the outer planets in our solar system.

Another indication of the impending doom will be a slow initial rise in the temperature of the Cosmic Microwave Background Radiation, which will no longer be detectable as microwaves as it shifts to the far infrared and then to visible light. It is estimated that the contraction will commence about 25 billion years from the present era when the Universe will have expanded to four times its present size. As the contraction accelerates, the environment everywhere will become increasingly difficult. The CMBR will rise to a temperature about 1000 times its present value of three degrees Kelvin and conditions will approach those of the Big Bang and then the end of all matter, space and time – the Big Crunch.

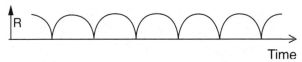

Fig. I.7 – The Cyclic Universe (K=+1)

But wait a minute! Some have speculated that the next Big Bang starts at the end of the Big Crunch, as shown in Fig. 1.7 which depicts the model of a cyclic or oscillating universe. This concept was eagerly supported by those who prefer a universe that does not require a First Cause and by those who have some fascination with Hindu mysticism. About 20 years ago (when there was considerable uncertainty about the mass density of the Universe), the idea of a cyclic universe seemed to me to be a verification of the Kaddish prayer – L'olam U'lolmay Olmaya – to this Universe and all the universes ever to be. At that time, I was not aware that the oscillating universe concept would dispense with the need for divine intervention, just as the earlier Steady State model accomplished the same objective for its supporters.

Another attractive feature of the cyclic universe concept was its denial of total oblivion; each new cycle is reborn, rising like the Phoenix from the ashes of the preceding universe. However, scientists like Richard Tolman over 50 years ago pointed to a number of flaws in the picture of Fig. 1.7. For example, conversion of matter into starlight via nuclear fusion (according to the accepted theory of stellar evolution) means that a particular cycle would approach its Big Crunch with more radiant energy and less matter than it had at its birth. The new universe would thus expand at a faster rate, reach a higher apogee and extend over a longer period of time than its predecessor. Each succeeding cycle becomes wider and bigger, as shown in Fig. 1.8, corresponding to decreasing values of k, the space-time curvature. When mass density drops below the critical value in a particular cycle, that universe is no longer closed and expansion reaches escape velocity, as shown in the seventh cycle of Fig. 1.8.

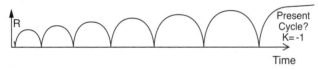

Fig. I.8 – Each Cycle expands as K decreases from +1

In our universe, the ratio of radiation (photons) to matter is large but not infinite. This is a very significant point, since it implies that there was a finite number of cycles prior to our present era. This deduction contradicts and demolishes one of the cherished features of an oscillating universe, stretching from minus to plus infinity in time – the lack of a need for a first cause. As shown in Fig. 1.8, there was a beginning at the first cycle, although not Genesis, which applies to our present cycle. So the question for our present world becomes, "Which of these cycles do we inhabit?"

The answer may lie in the latest observations, beginning in 1998, which indicate that expansion is not slowing down but is rather accelerating. There appears to be some source of dark energy that keeps increasing the velocity of expansion, just like a rocket which kicked in a propulsion system after coasting for some time. If correct, then the Universe appears to be headed for the "heat death" of Helmholtz and not the Big Crunch. This discovery, more than 70 years after Einstein introduced his cosmological constant and subsequently discarded it (section B), raises a number of questions as to its physical basis.

Apparently, we are in now in that last cycle, going towards icy extinction some trillions of years hence. As one popular newspaper article headlined, "T.S. Eliot wins and Robert Frost loses."

At this point, the astute reader may have perceived the suggestive analogy between the limited cyclic universes of Fig. 1.8 and the sabbatical cycles of Rabbi Isaac of Akko and the *Sefer ha-Temunah*. Since the collective memories of the preceeding cycles would have been obliterated by each bang and crunch, it is easy to speculate. Combining the latest data on expansion with the concept of these sabbatical cycles, we conclude that there were indeed six previous cycles and our Universe is in the seventh and final cycle. However, the age of *this* universe does not depend on prior cycles in which "Worlds were created and destroyed" as appears in the sources mentioned in Section E. Even if each succeeding cycle is larger and wider than its predecessors, one should ask how the authors of the *Sefer ha-Temunah* were able to formulate the basic idea of a cyclic universe more than 2000 years ago.

In his 1971 book on the Big Bang, "The First Three Minutes", Steven Weinberg ended with an epilogue entitled "The Prospect Ahead." He concludes that "whichever model is correct, there is not much hope in any of these."

More than a decade later, Andre Linde may have provided a ray of hope for our descendants, some trillions of years into the distant future. In a paper published in *Physics Letters* with the evocative title "Life after Inflation," he points out that our Universe is just one of a vast number of bubble universes which are constantly forming, inflating, flourishing and expiring (Ref. 1.19). The size of a typical bubble can be quite immense, as large as or even larger than our bubble. Each bubble may be separated from the other bubbles by insurmountable distances. On the other hand, should our descendants be fortunate enough to colonize a galaxy at the "edge" of our bubble, it may be possible with trillion-year technology to visualize an inter-bubble fleet of Noah's Arks which will bring them to a youthful universe in formation.

A favorite mode of interstellar space travel for science fiction writers has recently become a subject of intensive study, speculation and discussion by serious scientists. Stimulated by advances in string theory (which postulates that as many as 26 dimensions may be necessary to provide a unified T.O.E. – Theory of Everything), a number of authors have written and published on the possibility of space and time travel using wormholes (Ref. 1.20). Except for our familiar space-time dimensions, the rest of the dimensions required by the mathematical descriptions of string theory reality are hidden and invisible. It is these hidden, curled up dimensions that make wormholes possible. Consider Fig. 1.9 from Ref. 1.20. Two types of wormholes are illustrated; in the first case, a wormhole connects one part of our universe to another part of the same cosmos. Five billion years hence, our descendants will need to have developed the technology to either find or construct a suitable wormhole by which to escape from the solar system (as our sun dies) and migrate to another within our galaxy or even beyond. This will be no mean task, since according to some calculations, the energy requirements are tremendous – about 10^{-19} GeV (GeV=one billion electron volts). The highest energy of our current Super-Collider machines are "only" 1000 GeV, about 16 orders of magnitude less than the estimated energy needed. There are many other formidable obstacles, including survivability of space crews and passengers, which will need to be addressed and overcome.

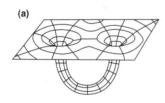
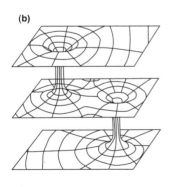

Fig. I.9 – Two types of wormholes
(a) Intergalactic connection in our universe
(b) Connections between parallel universes

The second set of wormholes is shown interconnecting parallel universes. As our own universe winds down to the Helmholtz heat death or approaches the Big Crunch (about 30 billion to almost a trillion years hence), another migration, this time to another universe, will be undertaken. It is fortuitous indeed that the magnitude of the difficulties to be surmounted escalates in proportion to the time remaining to develop the technological breakthroughs needed to ensure the survival of the species.

In my world-view, such a future clearly becomes a fulfillment of the Biblical and liturgical response in the Kaddish, "May His great name be blessed in this universe and all the many universes to be."

David Medved

Appendix 1.6 – Leviticus XXV and the Seven Sabbatical Cycles
The laws of *Shmita* (allowing the land to lie fallow every seventh year) and Jubilee (50th year) are stated as follows: *"And thou shalt number seven Sabbaths of years unto thee, seven times seven years and the period of the seven Sabbaths of years shall be unto thee forty and nine years. Then shalt thou cause the cornet of the Jubilee to sound on the tenth day of the seventh month; in the day of expiation* [namely, *Yom Kippur*] *shall ye make the cornet pass throughout all your land. And ye shall sanctify the fiftieth year and proclaim liberty throughout the land and all the inhabitants thereof..."*

There are several observations on this passage:

1. The last phrase in XXV:10 should sound familiar; these words are etched in perpetuity on the Liberty Bell in Philadelphia. So much for those atheists who seek to prove that America's Founding Fathers were pure secularists and who keep trying to remove such phrases as "In God We Trust" from U.S. currency and "one nation under God" from the Pledge of Allegiance.
2. Rashi has pointed out a remarkable "coincidence" in his commentary on Leviticus XXVI:32–35 (predicting the destruction of the First Temple and the first dispersal of the Jews). Line 34 states: "Then shall the land enjoy her Sabbaths, as long as it lieth desolate and ye be in your enemies' land; even then shall the land rest and enjoy her Sabbaths." Rashi points out that the seventy years of the Babylonian exile coincide *exactly* with the number of *Shmita* and Jubilee years which were *not* observed by the nation (the Ten Tribes of Israel and Judah-Benjamin) over a 430-year period. Even the most secular of authorities would agree that the Book of Leviticus was written well before the Babylonian exile.
3. There is an interesting progression in the concept of the seven periods of time:
    a. Seven days in a week with the seventh day as a day of rest.
    b. Seven weeks of the Omer ("Count") between Passover

and Shavuot (Pentecost), culminating in the holiday of Shavuot on the fiftieth day
c. Seven years in the *Shmita* cycle with the seventh year as a year of rest for the land.
d. Seven times seven *Shmita* years leading toward Jubilee, the 50th year as a celebration for the land.
e. Seven thousand years in a Sabbatical cycle with the seven thousandth year as an undefined "dream time".
f. Seven Sabbatical cycles

References - Chapter 1

1.1 Leon R. Kass, "Science, Religion and the Human Future," *Commentary Magazine*, 123: 39 (April, 2007)

1.2 Commentary of the Ramban (Nachmanides), Genesis ch. VI, verse 33

1.3 Rob Yule (St. Albans Presbyterian Church), "Forming the Universe-Genesis 1.2" [http://www.stalbans.org.nz/teachings/rob_yule/creation/form_uni.htm]

1.4 R.A. Alpher and R.C. Herman, "Evolution of the Universe", *Nature*, 162: 774–775 (1948)

1.5 A.A. Penzias and R.W. Wilson, "A Measurement of Excess Antenna Temperature at 4080 Mc/s", *Astrophysical Journal*, 142:419–421 (July, 1965)

1.6 R.H. Dicke, et. al., "Cosmic Black-Body Radiation", *Astrophysical Journal*, 142:414–419 (July, 1965)

1.7 The Maharal, *Tifereth Israel*, chapter 26 (*"Since the time and space constitute a single entity as is well known to those who understand"*)

1.8 Edgar Allan Poe, "Eureka, a Prose Poem", (1848) [This entire work can be accessed at *http://eserver.org/books/poe/eureka.html*]

1.9 Sam Aronson. et. al., "Researchers: Early Universe Liquid-Like", *Meeting of the American Physical Society*, Tampa, Florida (April 18, 2005)

1.10 Michael Oard, "Missing Anti-Matter Challenges the Big Bang Theory", *Answers in Genesis*, Vol. 12, Issue 3 (Dec., 1998)

1.11 Dr. Janet Conrad and Dr. William C. Lewis, "How Did the Universe Survive the Big Bang? (In this experiment, clues remain elusive)", *Presentation to an audience at Fermilab* (NY Times, April 12, 2007)

1.12 Steve Naftilan (Claremont Colleges), Commentary on ScientificAmerican.com – posted on Oct. 21, 1999

1.13 "Anthological commentaries from the Talmud – Midrashic and Rabbinic Sources on Genesis", pp. 44–47, *Art Scroll Tanach* series, Vol I, Mesorah Publications Ltd., (June, 1977)

1.14 F.J. Carrera, A.C. Fabian, & X. Barcons, "Soft x-Ray Background

Fluctuations and Large Scale Structure in the Universe", *Royal Astronomical Society Monthly Notices*, Issue 4 Vol. 285:820–830 (March, 1997)

1.15 A. Einstein, "On an Heuristic Viewpoint Concerning the Production and the Transformation of Light", *Annalen der Physik* (1905) – in German [English translation available in the *American Journal of Physics*, Vol. 33, May, 1965 by A.B. Arons and M.B. Peppard]

1.16 E.G. Freudenstein, *Intercom XII* (2), 1971 (published by the Association of Orthodox Jewish Scientists)

1.17 Aryeh Kaplan, *Immortality, Resurrection and the Age of the Universe*, published by the Association of Orthodox Jewish Scientists, 1993

1.18 Paul Davies, *The Last Three Minutes*, Basic Books (1994)

1.19 A.D. Linde, "Life after Inflation", *Physics Letters,* Series B, Vol. 211:29 (1988)

1.20 Michio Kaku, *Hyperspace – A Scientific Odyssey Through Parallel Universes, Time Warps and the 10th Dimension*, Anchor Books (1995)

Chapter 11:

The Music of the Celestial Spheres (Astronomy and the Bible)

The first five lines of Psalm 19 constitute a recurrent theme of this chapter. The Heavens are depicted as testifying to the glory and magnificence of Hashem's creation in "The Music of the Spheres." Several questions are raised on a proper understanding of their message and the need to resolve some apparent contradictions, especially in the phrase "בלי נשמע קולם" – *"their sound is unheard."*

There are varying opinions of our Sages and Commentators on the meaning of these five sentences. Rashi, et.al. believed that the heavens do not speak, but stimulate mankind to gaze in awe at these wonders of God's creation. Rambam, on the other hand, writes that "...the Psalmist really means to describe...what the spheres actually do and not what man actually thinks of them." Ibn Ezra says that one must be well versed in astronomy to understand the celestial magnificence of Psalm 19. William Shakespeare writes that the heavens speak to the inner soul of man.

500 years after King David, the Pythagoreans postulated that the spheres emit musical sounds but at frequencies outside the range of human hearing. We trace the evolution of these concepts by historians, philosophers, poets and composers through the following 25 centuries, culminating in the latest discoveries in modern science. Pulsars, gravitational waves, acoustic oscillations in the early universe, solar ultra-sound and other phenomena are cited as confirmations of the secrets embedded in the Bible.

1) Psalm 19 (King David, ca. 1000 BCE)

למנצח מזמור לדוד השמים מספרים כבוד אל ומעשה ידיו מגיד הרקיע;
יום ליום יביע אמר ולילה לילילה יחוה דעת; אין אמר ואין דברים בלי נשמע
קולם; בכל הארץ יצה קום ובקצה תבל מליהם

The following translation is the author's synthesis of those found in the Jerusalem Bible and the Artscroll Siddur:

> "¹ To the Conductor: a Psalm of David. ² The Heavens declare the glory of God; and the firmament proclaims his handiwork. ³ Day to day brings expressions of praise, and night following night bespeaks wisdom. ⁴ There is no speech nor are there words, their sound is unheard. ⁵ Their line goes forth throughout the earth and their words reach the farthest ends of the land..."

Nature is depicted as singing in testimony to the glory of God. Each component of the Universe performs its designed function in harmony with all the other parts of Creation. Although used as an introduction to additional Psalms sung by the Levites in the Temple, the word למנצח (to the Conductor) may have very special meaning here. It has also been translated as "to the Chief Musician."

In this Psalm, the Universe is viewed as a symphony orchestra under the baton of a Master Conductor, playing hymns of praise for the continuous wonders of Creation. The visible planets and constellations all bear witness to a glorious Universe of staggering dimensions, brought into being during the six days of Creation (Genesis 1). The young shepherd David, tending his father's flock in the hills of Judea 3000 years ago, contemplated a sky unobscured by pollution and city lights.

Avraham Ibn Ezra comments that unless one is well versed in astronomy, one cannot really appreciate the celestial magnificence of this Psalm (Ref. 11.1). Further on, in the same reference, he writes that seeing and hearing have two components, one physical and one spiritual. Our eyes and ears detect the physical component, whereas man's inner soul is responsive to the spiritual message. The Radak, almost one hundred years later, explicitly states that: "the inner soul of man clearly discerns their message" and that "their words are more eloquent than the spoken word..." Shakespeare may have been more familiar with the writings of these commentators on Scripture than most people realize (see next section).

It is easy to be swept up by the lyrical beauty of these lines, but a few questions need to be addressed from an analytical perspective:

a. Line 4 appears to be in contradiction with the preceding lines 2 and 3.
b. In line 5, the phrase "their words reach the farthest ends of the land" reverses the contradiction of line 4.
c. Unlike light, sound cannot propagate through the vacuum of space between our Earth and the luminaries.

Most of the commentators, including Rashi, Radak, Malbim and others, try to resolve the dilemma by explaining, "Are the heavens capable of speech? No. But they can stimulate men to articulate the praises of God." If this was the sole opinion,, it would be the end of our story. However, Rambam (Maimonides) vigorously disputes this view, writing in his Guide to the Perplexed (Ref. 11.2):

"It is a great error to think that this [Psalm 19] is a mere Figure of speech...the Psalmist really means to describe the heavens' own doing, or, in other words, what the spheres actually do and not what man thinks of them..." Later, in Chapter VIII, Rambam discusses the mighty and fearful sounds emitted by the sun, moon and stars, considering their great sizes and velocities. He then states "The Pythagoreans believed that the sounds were pleasant...and they also explained why these mighty and tremendous sounds are not heard by us."

2) From Pythagoras (Sixth Century BCE) to Holst (1930)

Pythagoras and his school taught that the heavens could be described as a structure of concentric spheres, each supporting a known planet, with the sun as the innermost sphere. This cosmic harmony both produced, and was in itself, music. The Pythagoreans believed that all matter emitted musical tones, but at frequencies outside the range of human hearing. Line 4 of Psalm 19, in stating "...their sound is unheard," may have influenced the Pythagoreans in their development of these concepts.

Most secular scholars do not agree with the Jewish tradition that King David wrote the psalms and some even question his existence. Their influence is so pervasive that every reference to the music of the spheres states unequivocally that Pythagoras was the first to introduce this concept. It is tempting to speculate that Jewish refugees, fleeing westward following the destruction of Solomon's Temple in 586 BCE, landed on the shores of Greece and transmitted their traditions to the early Pythagoreans.

Pliny the Elder (Gaius Plinius Secundus, 23–79) was one of the foremost intellects of the Roman world, who made the whole realm of nature and history the central theme of his life and writing. He gives a description of the musical intervals corresponding to the radii of the planetary orbits (Ref. 11.3), and conceives an analogy between the strings of a lyre and the distances from the Earth to the planets. For example, the Moon, as the closest heavenly body, would be represented by the shortest string, whereas Saturn would be the longest string on this celestial lyre.

The frequency of a vibrating string is inversely proportional to its length and varies according to the square root of its weight or tension. The dimensions of a piano are reasonable only because the higher notes use heavier strings. An instrument called the monochord was familiar to the Pythagoreans and is still found in many acoustics laboratories (see Fig II.1 from Ref. II.4). The monochord may have been the inspiration for Pliny's celestial lyre, wherein the Earth is the rigid point A, the length of the wire B to C represents the distance from Earth to the moon, and the weight, W, is the mass of the moon.

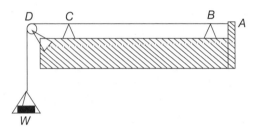

Fig II.1 – The Monochord: Variable bridges, B–C and the suspended weight, W, determine the frequency

We will encounter Pliny's writings and theories in our discussions of the constellation Lyrae (the Harp) later in this chapter and several more times in Chapter IV on Chemistry and the Bible.

Pliny died in the eruption of Vesuvius after launching ships from Misenum across the Bay of Naples to Pompeii (August 24, 79), against the better judgement of his commanders. He was a close friend of the Emperor Vespasian, who had appointed him an Admiral of the Roman fleet. Motivated by his scientific curiosity and a desire to observe and record the awesome phenomenon of a volcano at its source, he insisted on a closer look. He had also hoped to rescue survivors and the priceless manuscripts of the great library of Pompeii. His nephew, Pliny the Younger, described these events in his letters to Tacitus, and a recent historical novel provides a vivid and dramatic account of Pliny's last moments (Ref. II.5). Volcanologists employ the term *plinian* when describing a very violent volcanic eruption and *ultra-plinian* for the most violent in history, such as Krakatoa in 1883.

Johannes Kepler (1571–1630), whose mathematical laws on the elliptical planetary orbits laid the groundwork for Newton, taught that these orbits were arranged like a musical progression. His contemporary William Shakespeare wrote, in "Merchant of Venice", "There's not the smallest orb which thou behold'st but in his motion like an angel sings" (Ref. 11.6). The master poet of the English language at first refers to the sounds of man-made music in Lorenzo's instructions to Stephano. Lorenzo then gives voice to the message of Psalm 19 and the commentaries of Ibn Ezra and Radak, i.e. the music of the spheres cannot be heard by mortal humans but only in harmony with our immortal souls.

LORENZO (TO JESSICA, DAUGHTER OF SHYLOCK)
    Sweet soul, let's in, and there expect their coming.
    And yet no matter: why should we go in?
    My friend Stephano, signify, I pray you,
    Within the house, your mistress is at hand;
    And bring your music forth into the air.

[Exit Stephano]
    How sweet the moonlight sleeps upon this bank!
    Here will we sit and let the sounds of music
    Creep in our ears: soft stillness and the night
    Become the touches of sweet harmony.
    Sit, Jessica. Look how the floor of heaven
    Is thick inlaid with patines of bright gold:
    There's not the smallest orb which thou behold'st
    But in his motion like an angel sings,
    Still quiring to the young-eyed cherubins;
    Such harmony is in immortal souls;
    But whilst this muddy vesture of decay
    Doth grossly close it in, we cannot hear it.

It is quite amazing that Shakespeare puts such sublime language in the mouth of a thief and seducer. Nevertheless, by a few strokes of his pen, he came close to resolving the questions raised by lines 2 to 5 of Psalm 19.

Gustav Holst composed "The Planets Suite" in 1916 in the midst of WWI and personally conducted it in 1930, shortly before the discovery of Pluto. The ancient seven visible bodies (the sun, moon and five planets) had already been increased by the telescopic discoveries of Uranus and Neptune. Apparently, Holst had studied mythology and astronomy before setting the planets to music (replacing the sun and the moon with Uranus and Neptune). The piece starts with the loud brash entry of Mars and ends with a mystical fadeout of Neptune, perhaps representing the mysterious edge of the solar system.

3) Bode's Law and Musical Octaves

Following the work of Kepler and Newton, Johann D. Titius (1766) developed an approximate mathematical relationship between the distances of the planetary orbits from the sun. In 1772, this work was published by Johann E. Bode and is known today as Bode's Law.

Consider the sequence: 0, 0.3, 0.6, 1.2, 2.4, …where each number is double the preceding one. The sequence is written in Astronomical Units (A.U.), where one A.U. is the average distance of the Earth from the Sun (approximately 150 million km.). Bode's Law states that assigning Mercury as the zeroth planet, and adding 0.4 A.U. to it and the succeeding orbits in the sequence, gives the results summarized in Table 11.1.

Table 11.1 – Bode's Law

Planet	Distance from sun (millions km)	Distance from sun (A.U.)	Predicted Distance	Sequence
Mercury	58	0.39	0.40	0
Venus	110	0.71	0.70	0.3
Earth	150	1.00	1.00	0.6
Mars	230	1.52	1.60	1.2

Asteroids	440	2.93	2.80	2.4
Jupiter	780	5.20	5.20	4.8
Saturn	1430	9.50	10.00	9.6
Uranus	2880	19.20	19.60	19.2
Neptune	4516	30.10	38.80	38.4
Pluto	5940	39.60	77.20	76.8

There is good accord between the predicted values and the actual distances (better than 5%), but the model breaks down when applied to Neptune and Pluto, discovered well after its publication. Most astronomers now consider it a fortuitous coincidence; however, at a meeting of the International Astronomical Union (Prague, August, 2006), Pluto was demoted from its planetary status and was tossed into the bin of dwarfs (44 to date).

Modern music (since the 18th century) is based on octaves, each containing 12 notes (seven major notes and five minor notes). Middle C (C_4), at a frequency equal to 261.6 cycles per second, is separated from C_5 in the next octave by a factor of two, i.e., $C_5 = 523.2$ cps. The lowest note sounded by a double bass or brass tuba is about 40 cps. The range of human hearing extends over approximately 10 octaves, from 20 cps., with the highest frequency at 20,480 cps. If we treat Earth as a special case (Ref. 11.7), the orbital distances of the inner planets up to Jupiter are analogous to the first four octaves of the tempered scale, i.e. each orbit is double that of the preceeding one.

The composer of Psalm 19, Pythagoras and all those who followed, although not cognizant of the quantitative relationships between planetary orbits and musical scales, had an intuitive grasp of the Music of the Spheres.

A Timeline summary of these poets, philosophers, musicians, scientists and Torah scholars is given in Appendix 11.1.

4) The Secrets of Science in Psalm 19

As noted in our discussion of the dilemmas posed by Psalm 19, sound waves at any frequency require a medium for their transmission. Since the vacuum of the cosmos cannot transmit sound waves, one might conjecture that "their words" (line 5 of the Psalm) are transmitted to the farthest ends of the land as modulated light of the luminaries, wherein the modulation frequency is within the acoustic range. In fact, modern astronomy is founded on a long history of observations on fluctuations in light and radio signals from a variety of sources. The Cepheid variables, known as the Candles of the Universe, exhibit slow periodic variations of their luminosity, as do the stars known as RR Lyrae in the constellation of Lyra (The Harp).

William Huggins, the founder of stellar spectroscopy, wrote, "Within this unraveled starlight exists a strange cryptography. Some of the rays may be blotted out, others may be enhanced in brilliancy. Their differences, countless in variety, form a code of signals...." (as quoted in Ref. 11.7, p. 119).

Pulsars

In 1967, Jocelyn Bell a graduate student working with Professor Antony Hewish on radio astronomy at Cambridge University, detected extra-terrestrial radio transmissions of periodic pulses from four different stellar sources. After a brief flurry of excitement, thinking that these signals originated from alien civilizations in our galaxy, it was determined that these pulses (called LGM for Little Green Men) actually came from spinning neutron stars, which are now believed to be the residue of a supernova explosion. At this writing, about 1500 RF (radio frequency) pulsars have been observed and catalogued, all at repetition rates within the audio range, to a maximal rate of 700 pulses per second.

A few of these spinning stars also emit optical pulses, x-rays and gamma-rays in sync with the RF, like the pulsar in the Crab Nebula. The supernova explosion, which created the Crab Nebula in the constellation Taurus in 1054, was recorded by astronomers in China and possibly by the Anasazi Indians in what is now Arizona/ New Mexico. The Chinese reported that its brightness was more than four times Venus and that this new star was visible during daylight hours for a three week period (the estimated distance to the Crab Nebula is 6500 light-years). The gamma-ray bursts were first observed by American satellites and were thought to be the result of clandestine Soviet tests in violation of the test ban on nuclear explosions in space.

A supernova results when a star with mass between eight and thirty solar masses has consumed all of its fuel and converted it to iron. Iron (Fe56) is the terminus of the fusion process since its nucleons have the highest binding energy of all the elements, so that fusion at this point requires a net energy input. As a result the star undergoes a gravitational collapse, imploding and blowing off much of its upper layers as a supernova. What remains will become a neutron star, white dwarf or black hole (depending on the residual mass). In the neutron star, the intense gravity has compacted the atoms to produce neutrons and intense magnetic fields, about one trillion times that of Earth. The rapid rotation in this immense magnetic field acts as a giant celestial dynamo that produces an extremely powerful electric field, which draws electrons and other ionized particles out of the star's crust. Fortuitously, the axis of the magnetic field is offset from the spin axis (Fig. II. 2), just as the weak magnetic field of the Earth is tilted relative to the axis of rotation. The source of the RF and other signals are the electrons and ions which move along the field lines.

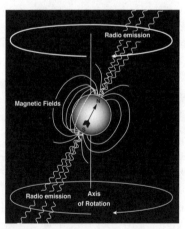

Fig II.2 – Physics of the pulsar – a rapidly rotating neutron star

The periodic RF pulses are analogous to the rotating beam at the top of a lighthouse as it sweeps through the foggy night. The period (the time between the pulses) of the 1500 pulsar transmissions varies up to a few thousandths of a second, corresponding to a modulation frequency right in the middle of our hearing range. These pulsating radio signals arriving at Earth 3000 years after the composition of Psalm 19 and from various parts of our galaxy certainly fit the descriptive eloquence of those first five lines.

Gravitational Waves

According to the Theory of General Relativity, published by Einstein in 1915, all objects in the Universe produce a distortion in the space-time fabric of their vicinity, just by being there. The degree of this "space-warp" varies with the mass of the object, as the heavier the body – the greater the warp. The earth maintains its orbit around the sun, since it is constrained to follow the geodesic curves in the space-time warp produced by the presence of the sun. The Einstein equations predicted that the motions of the celestial bodies, or any violent occurrence such as a supernova explosion, would produce a gravitational wave. Just as the electrons moving to and fro in a radio antenna produce an electromagnetic wave which can be detected by distant receivers, these gravitational waves are generated by the motions of the heavenly bodies through the fabric of the cosmos. As sound waves are oscillations of the medium (like air) and also propagate through the same medium, gravitational waves are oscillations of the space-time continuum and also travel through space-time.

The amplitude of these waves depends on several factors:

A. the mass of the source
B. the acceleration of the source
C. the asymmetry of the source, i.e. its 'lumpiness', defined by scientists as its quadrupole moment
D. the distance of the source

For more than 50 years following the publication of Einstein's theory there was little interest in this subject, and we have only indirect evidence of the existence of these waves from observations of binary pulsars (Fig. 11.3). These are rapidly spinning neutron stars which orbit around each other, similar to a cosmic rotating dumbbell.

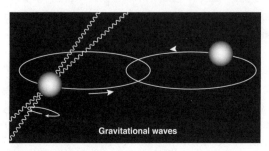

Fig II.3 – Binary neutron stars in tight orbits emit gravitational waves

In 1974, Hulse and Taylor observed that the orbital period of the binary pulsar PSR B1913+16 was decreasing, indicating a loss of energy by this binary system (Ref. 11.8) They concluded that this energy loss was due to the emission of gravitational waves, and employed the Einstein Equations to calculate the decay of the orbital period. The calculations agreed with the observations to better than one part in one thousand. More recently (in 2005), observations on another binary system, PSR J0737–3037, again confirmed the predictions of General Relativity on the energy loss resulting from the emission of gravitational waves. These observations showed that the orbit of this binary pair was decreasing at a rate of 7 mm. per terrestrial day, confirming the calculated decrease.

Objects in our solar system will radiate only small amounts of gravitational wave energy, since even gas giants like Jupiter and our sun produce relatively small curvatures in space-time, and this curvature is changing slowly. These bodies are also more or less spherically symmetric. Even though the distances to the stars are many orders of magnitude greater than to our companions in the solar system, it is expected that the first detection of gravitational waves will be from extra-solar sources. A number of experiments using laser interferometers with long light paths, such as LIGO and the space-based LISA, are in process. LIGO is being set up in two different locations with 4 km arms, with the amount of change in length expected to be 10^{-16} cm.

Some scientists have conjectured that it might be possible to detect the primordial gravitational waves produced by the expansion of the early Universe. This would allow us a peek into its Dark Age – the time before the photons (which later became the Cosmic Microwave Background Radiation, or CMBR) were able to escape from the embrace of the ionized plasma. If detectable, these waves would provide the first direct evidence of the conditions immediately following Genesis.

In order to improve the detection efficiency of LIGO and other instruments, scientists have calculated the expected frequency of gravitational waves which are produced by supernovae, binary pulsars, and black hole collisions. Brian Greene in his recent book, "The Fabric of the Cosmos," writes:

"Curiously, the calculations show that some gravitational wave frequencies should be in the range of a few thousand cycles per second; if these were sound waves, they'd be right in the range of human audibility" (Ref. 11.9).

In a paper recently submitted for publication (June, 2006), N. Christensen writes that LIGO has achieved its initial target sensitivity and that a world-wide detection network (VIRGO in Italy, GEO in Germany and TAMA in Japan) is also coming on line. All of these ground-based laser interferometers are tuned to be sensitive to gravitational wave radiation in the frequency band from 40 Hz. to 8 kHz, spanning almost exactly the range of human hearing (Ref. 11.10). It's an amazing "coincidence" that the two very different heavenly phenomena (pulsar RF modulation and gravitational waves) both have frequencies which would fall right in the range of human hearing, but which we cannot hear, just as implied in the first five lines of Psalm 19. Is this another case in which the secrets of 21st century science are indeed embedded in the Bible?

Cosmic Sound

In addition to modulated wave phenomena which reach Earth in the acoustic range, there are several "localized" effects which have been dubbed by the scientists studying them as "cosmic sound."

BAO *(Baryon Acoustic Oscillations)*

A baryon is an elementary particle of large mass (such as a proton or neutron), held together by a combination of three quarks In this first decade of the 21st century, some remarkable measurements on the fine structure of the cosmic microwave background radiation (CMBR) found acoustic peaks in this "surface of last scattering," According to present theory, about 370,000 years after the Big Bang, the primordial plasma soup of photons, baryons and electrons had cooled to a temperature of 3000 degees Kelvin, at which point the equilibrium shifted to formation of neutral hydrogen atoms. The photons at this temperature no longer had sufficient energy to ionize these neutral atoms and were able to escape the plasma trap, thus ending the Dark Age of the Universe (compare Genesis 1:2,3 and Chagiga 12a). Today, the CMBR has cooled to a temperature of 2.7 degrees K, as a result of redshift connected to the expansion of the Universe.

Two NASA satellites, COBE and WMAP, launched in 1989 and 2001 respectively, have transformed experimental cosmology into an exact science by the remarkable precision of their data on the structure of the CMBR. The 2006 Nobel Prize for Physics was awarded to John C. Mather and George F. Smoot, the project leaders on COBE, in recognition of their supervision of more than 1000 scientists and engineers and their personal involvement in the measurement and analysis of the black-body spectrum and the anisotropies in the temperature distribution of the CMBR (to an accuracy of one hundred-thousandth of a degree). The WMAP satellite improved the resolution to one-millionth of a degree.

The baryon acoustic oscillations were longitudinal sound waves, roiling through the early plasma and increasing in wavelength and decreasing in frequency as the Universe expanded. Their existence was first predicted by Peebles and Yu in 1970 (Ref. 11.11) and independently by Sunyaev and Yakov Zeldovich in the same year (Ref. 11.12). These sound waves resulted from the tug of war between gravity and photon pressure. The overly dense regions became the galaxy clusters and the under-dense portions are the great voids between the galaxies. Since the wavelengths of these sound waves at the end of the Dark Age extended over a universe whose dimensions were 370,000 light years, their frequency would be many orders of magnitude below the level of human hearing. The analysis of the acoustic peaks in the anisotropy power spectrum of the CMBR has become one of the most powerful cosmological tools in our studies of the bouncing baby Universe.

Following the mapping of the density fluctuations in the two-dimensional CMBR, Prof. Daniel Eisenstein at the University of Arizona and a legion of co-investigators carried out a complementary three-dimensional survey of galactic distributions. In a paper published in 2005 (with more than 30 authors from about 15 institutions), they presented the results and analysis on a large sample extending over 46,748 galaxies (Ref. 11.13). Designated SDSS (Sloan Digital Sky Survey), it covered a volume equal to a cube four billion light-years on each side (9% of the sky), and took more than five years to complete, using a dedicated state-of-the-art telescope at the Apache Point Observatory, New Mexico. Their data at low redshift, combined with the CMBR acoustic scale, have confirmed several basic tenets of cosmological structure formation theory. Daniel Eisenstein writes, "By demonstrating that this imprint of the early universe (namely, the CMBR acoustic anisotropies) survives the 13 billion years of growth to exist in the clustering of the galaxies today, we have confirmed the theory for the gravitational growth of structure" (Ref. 11.14).

Solar Ultrasound
In a communication published in Astrophysical Journal Letters, researchers at the Southwest Reseach Institute (SwRI) reported that the sun's atmosphere contains evidence of ultrasonic waves at a frequency of 100 millihertz (100mHz), with a 10 second period (Ref. 11.15). The signature of these sound waves was discovered by

analysis of images recorded by the ultraviolet telescope mounted on the TRACE (Transition Region And Coronal Explorer) spacecraft, launched into earth orbit in 1998). Craig DeForest of SwRI was quoted as saying, "These ripples seem to be carrying about one kilowatt of power per square meter on the surface of the sun – that is similar to the sonic energy you might find coming out of the speakers at a rock concert. Very loud!"

This statement, talking about a discovery almost one thousand years after Rambam, should be compared with Chapter VIII, Part II of his Guide to the Perplexed (see Ref. 11.2), where he writes, "Thus, our sages describe the greatness of the sound produced by the sun in the daily circuit of its orbit."

At 100 mHz, these sound waves are approximately 300 times below the low frequency threshold of the human ear. This doesn't deter popular science writers from dubbing this solar ultrasound the "Bass Note in Music of the Spheres" (Ref. 11.16).

Scientists hope that a more detailed study of the solar ultrasound will help them to elucidate the source of the sun's extra heat. The solar surface temperature is about 6000 degrees Celsius and is responsible for the spectrum of the black body radiation we receive on Earth. The middle solar atmosphere (chromosphere) is considerably hotter at 100,000 degrees and the solar corona is a sizzling one million degrees. It is believed that further analysis of the ultraviolet carrier of this sonic information will finally enable us to decipher the mystery of this extra heat, which has baffled the scientific community for more than 50 years (Ref. 11.17).

5) Conclusions

Recent discoveries in astronomy have strikingly confirmed the basic thesis and meaning of the first five lines in Psalm 19:

a. The celestial spheres do indeed send us sonic coded messages whose "sound is unheard." Like most of the ancient world, the Pythagoreans were not aware that sound waves cannot propagate in the vacuum of empty space. Instead, they postulated that the planets and stars were emitting musical sounds outside the range of human hearing.

b. On the other hand, acoustic waves do reach Earth as modulated information on a variety of carriers, including RF (radio frequency), optical and ultraviolet rays, gamma rays, x-rays and the fabric of space-time.

As the final editing on this chapter had just been completed, the author received an email from *Scientific American.com* with a most relevant article entitled, "Strange but True, Black Holes Sing," dated October 18, 2007. Authored by Robynne Boyd, it describes sound waves surrounding a massive black hole in the heart of the Perseus galaxy cluster. The sound waves were detected by variations in the x-ray emissions from the halo of gas surrounding the black hole. The source is designated as a cosmic drum producing concentric ripples of brighter and fainter gas with a period of about 10 million years. The frequency turned out to be the lowest celestial music yet observed, at 57 octaves below middle C (about 3×10 to minus 15 Hz). The article concludes with a discussion of other cosmic sounds like the BAO and solar ultrasound.

Table 11 summarizes the currently available data, listing period, frequency and observation method. Two of the phenomena exhibit frequencies which fall within human audio range, while the other four can be defined as ultrasonic, or below our hearing range.

Table 11: Frequencies of Celestial Phenomena

Phenomenon	Period	Freq. Range	Carrier/Observation
Cepheid Variables	3 to 30 days	0.3 to 3 µHz	Optical (telescope)
Rr Lyrae Stars	1 to 24 hrs	10 to 300 µHz	Optical (telescope)
Solar Ultrasound	10 sec	100 mHz	Ultra-violet (TRACE)
Baryon Acoustic Osc.	-	10-26 Hz	RF and optical (CMBR anisotropies and galactic distributions)
Pulsars	1.4 msec.	700 pps max.	RF primarily, plus some optical, gamma rays and x-rays
Gravitational Waves	25 msec to 125 usec	40 Hz to 8 kHz	Fabric of Space-time (LIGO)

Appendix 11.1 – God of Wonders

> In February of 2006, my sons gave me a most memorable birthday gift – a full week with them at the Rancho La Puerta resort in Tecate, Mexico. In response to an invitaion from Pastor Jim Garlow of the La Mesa Skyline Church, we drove the short distance one evening after an unforgettable border crossing. We were greeted by an enthusiastic crowd of several thousand congregants participating with us in an inspiring and even emotionally charged discussion, punctuated by stirring music from a great band. As part of my address to the audience, I presented a brief description of the ideas and material on the Music of the Celestial Spheres. The band and their singers then followed with a rousing rendition of a most appropriate original song as follows – what an uplifting moment!

God of Wonders
Lord of all creation
Of water, earth, and sky
The heavens are Your tabernacle
Glory to the Lord on high

God of wonders beyond our galaxy
You are holy, holy
The universe declares Your majesty
You are holy, holy

Lord of heaven and earth
Lord of heaven and earth

Early in the morning
I will celebrate the light
And as I stumble in the darkness
I will call Your name by night

God of wonders beyond our galaxy

Hidden Light: Science Secrets of the Bible

You are holy, holy
The universe declares Your majesty
You are holy, holy

Lord of heaven and earth
Lord of heaven and earth

Hallelujah! to the Lord of heaven and earth

God of Wonders – Steve Hindalong & Marc Byrd
2000 Meaux Mercy/Storm Boy Music

Appendix 11.2 – Timeline of the Concept

	Dates
David ben Yishai (King of Israel)	ca. 1000 BCE
Pythagoras (Greek Mathematician & Philosopher)	572–497 BCE
Pliny the Elder (Gaius Plinius Secundus) (Roman Author & Philosopher)	23–79
Rashi (Rav Shlomo Yitzhaki) (Bible & Talmud Commentator)	1040–1105
Avraham Ibn Ezra (Doctor & Philosopher)	1089–1167
Rambam (Rav Moshe ben Maimon) (Doctor & Philosopher)	1135–1204
Radak (Rav David Kimchi) (Bible Commentator)	1160–1235
William Shakespeare (British Poet & Playwright)	1564–1616
Johannes Kepler (German Mathematician & Astronomer)	1571–1630
Johann E. Bode (German Astronomer)	1747–1826
Gustav Holst (British Composer)	1874–1934

References – Chapter 11

11.1 Ibn Ezra on Tehilim 19 (as appears in מקראות גדולות (Mikraot G'dolot), Volume 3 of נ"ך, p. 11)

11.2 Rambam, Guide to the Perplexed, Part II, Chapter v, Friedlander Translation (Arabic to English), 1904

11.3 Pliny the Elder, Naturalis Historia xx, p. 84

11.4 Sir James Jeans, Science and Music, Cambridge Univ. Press, 1937 (also available by Dover Books), p. 62

11.5 Robert Harris, Pompeii, A Novel, Ballantine Books, 2003

11.6 William Shakespeare, "Merchant of Venice", Act v, Scene 1

11.7 Guillermo Gonzalez and Jay. W. Richards, The Privileged Planet, Regnery Publishing Inc., 2004

11.8 R.A. Hulse and J.H. Taylor, "A High Sensitivity Pulsar Survey," Astrophysical Journal, Vol. 191, part 2 (1974)

11.9 Brian Greene, The Fabric of the Cosmos, Vintage Books, 2005, p. 422

11.10 N. Christensen, "Searching for Gravitational Waves," paper submitted June, 2006 [see www.ligo.caltech.edu]

11.11 P.J.E Peebles and J.T. Yu, Astrophysical Journal, Vol. 162:815 (1970)

11.12 Rasheed Sunyaev and Yakov B. Zel'dovich, Astrophysics and Space Science, Vol. 7:3 (1970)

11.13 D.J. Eisenstein, et. al., "Detection of the Baryon Acoustic Peak in the Large-Scale Correlation Function of SDSS Luminous Red Galaxies," Astrophysical Journal Vol. 633: 560 (2005)

11.14. Daniel J. Eisenstein, "Dark Energy and Cosmic Sound", Steward Observatory, University of Arizona, 2005

11.15. Craig DeForest et. al., Astrophysical Journal Letters, Dec. 10, 2004

11.16. Bill Christensen, "Solar Ultrasound – Bass Note in Music of the Spheres," www.Technovelgy.com, Dec. 12, 2004

11.17 "Solar ultrasound waves provide clues about decades-old mysteries": Press Release from Southwest Research Institute (Dec.10, 2004) – available on www.spaceref.com

Chapter III
Mathematics and the Bible: Pi (π) in the Bible

A simple sentence in the first Book of Kings, describing the Great Basin in Solomon's Temple, has stimulated a great amount of interest, hilarity and anti-religious invective. Almost four million entries pop up when one does a search on Google for "pi in the bible," all dealing with VII:23 in 1 Kings. The verse merely states that the diameter of the Great Basin (also known as the Great Sea) was 10 cubits across and its circumference was 30 cubits, giving a value of pi = 3, whereas the neighboring civilizations of Egypt and Babylon (one thousand years earlier) had been using a significantly more accurate value than these seemingly primitive Hebrews.

This chapter begins with an historical review, describing the various methods employed to obtain more accurate values for Pi. Section 2 shows some amusing examples of criticism against the people of the Book and their God for this mistake. On the other side of the argument (Section 3) are a number of convoluted mathematical treatments attempting to rationalize this "error."

Section 4 proceeds to show in detail how our early exegetes were able to demonstrate that the revealed value of pi in this sentence was actually accurate to five significant Figures – a precision only exceeded by the work of Zu Chongzi in China, one thousand years later.

> "The Temple of Solomon was the most important embodiment of a future extramundane reality, a blueprint of heaven; to ascertain every last fact about it was one of the highest forms of knowledge, for here was the ultimate truth of God's Kingdom expressed in physical terms"
> – Sir Isaac Newton

1) A brief history of Pi

The constant ratio between the circumference of a circle (C) and its diameter (D) has fascinated mankind for more than 4000 years. This ratio is Pi.

Most astonishing is the current interest in two fairly obscure sentences in the Bible – 1) Kings I: VII:23 and 2) Chronicles II: IV:2.

1. Kings I: VII: 23:
   "And he made the molten (cast) sea 10 cubits from brim to brim: round all in compass, and its height was five cubits and a line of 30 cubits did circle it round about."

2. Chronicles II: IV:2
   "...he made the molten sea of 10 cubits from brim to brim, round in compass and five cubits its height and a line of 30 cubits did circle it round about."

The "he" in the quotes above is Huram of Tyre, a famed worker in metal (see Appendix III.1).

As of October 7, 2007, there were 3,990,000 entries in Google [up from November 15, 2005, when there were 2,600,000] on the topic of "Pi in the Bible." Many of these entries are divided between those who attack the ignorance of the ancient Hebrews (and their God who didn't know the correct value of Pi) and a large number of apologists who use intricate mathematical gymnastics to prove otherwise. Almost lost in the heat of this battle are several articles which explore the original Hebrew text, and, by using some fundamental rules of Biblical exegesis, have come up with a more accurate value to five significant Figures (see Section 4). The earliest records for the values of Pi are found in the writings of the Babylonians and the Egyptians, around 2000 BCE (the Rhind papyrus in Egypt and a Babylonian cuneiform tablet at the British Museum). The errors were less than 1% (see Table 1) and went uncontested for more than 1000 years.

Hidden Light: Science Secrets of the Bible

Table 1: The Values of Pi through History

Approx. Year	Culture	Source	Value	Difference from 3.1415926	% Error
2000 BCE	Babylonian	?	$3 \frac{1}{8} = 3.125$	-0.0165	0.5
2000–1650 BCE	Egyptian	Ahmes, the Rind Papyrus	$3 \frac{13}{81} = 3.1605$	+0.0185	0.6
1200 BCE	Chinese/Hindu	?	3.0	-0.14159	4.5
900 BCE to 610 BCE	Hebrew	Book of Kings-Jeremiah?	3.14150943	0.0000832	0.0026
250 BCE	Greek	Archimedes	3.14119	+0.00031	0.009
27 BCE	Roman	Marcus Vitruvius	$2\frac{5}{8} = 3.125$ (also used 4)	-0.0165	0.5
139	Chinese	Chang Hong	$\sqrt{10} = 3.1623$	+0.0207	0.02668
150	Greco-Egyptian	Ptolemy	$377/120 = 3.14166$	+0.000067	0.0021
450	Chinese	Zu Chongzhi	$355/113 = 3.1415929$	3×10^{-7}	8×10^{-6}
1400	Hindu	Madhava	to 11 places	-	-

The entry for Jeremiah will be discussed in detail in Section 4. Its accuracy was only equaled or exceeded by Ptolemy (150 CE) and Zu Chongzhi (450 CE) almost a millennium later. The calculation by Zu Chongzhi was the most accurate for another 1000 years until

Madhava carried out a calculation correct to 11 decimal places (1400 CE).

Almost all of these early formulations used the polygon method developed by Archimedes (around 250 BCE). Consider Fig. III.1. The Figure shows a circle of radius one unit (1) with a circumscribed

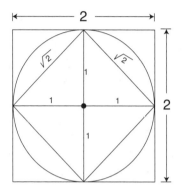

Fig III.1 – Archimedes Method

square whose sides equal the diameter of the circle and an inscribed square whose sides are √2 units (by the Pythagorean Theorem). The area of the circumscribed square is 4 units and the area of the inscribed square is thus 2 units (from $A=\pi r^2$). Therefore, a first approximation to π can be written as an inequality: 2<π<4

Now repeat the exercise with an 8-sided octagon and calculate the resulting areas. As the number of sides (n) continues to increase, the area of the outside n-gon shrinks and the area of the inside polygon grows, as given in Table II.

Table II – Area of polygons as a function of the number of sides

No of sides	Inner n-gon Area	Outer n-gon Area	Approximate value of π (on average)
4	2	4	3
8	2.8284	3.3137	3.07
32	3.1214	3.1517	3.136
96	3.1408 (3 10/71)	3.1429 (3 1/7)	3.1419

Archimedes' brilliant and innovative method is designated "the method of exhaustion," not necessarily referring to the physical state of the mathematician doing the calculation, but rather to the way the growing inner polygon exhausts the area of the host circle.

Euclid, who lived a century before Archimedes, never attempted to estimate the value of Pi. As a pure mathematician, he was only interested in the logical proofs of his theorems, such as "The area of a circle divided by the square of its radius gives the same numerical result [π] no matter how large a circle" [Elements XII; Proposition 2 – "Circles are to one another as the squares on their diameters"].

Contrast this approach with that of the Roman architect, Marcus Vitruvius Pollio (25 BCE), whom Herod the Great brought to Jerusalem to work on the expansion of the Second Temple. He determined Pi by measuring the distance a wheel, of a given diameter, moved through one revolution.

As occurred in many scientific disciplines, the lead in Pi shifted from the East back to Western Europe at the reawakening of the Renaissance. Van Ceulen set a new record in 1596 by achieving a value correct to 35 decimal places, and the race was on its way using mathematical formulations. By the dawn of the computer age, Ferguson had calculated Pi to 808 decimal places (1947). Using algorithms and computers, π has been calculated to more than 10 billion places. It will never stop since it is an irrational number. A rational number can be defined as the quotient of two integers (like ⅓). Note that ⅓ also has a never-ending decimal (0.333....) but it is periodic. Every irrational number has a decimal expression, which neither terminates nor is periodic.

A history of Pi would not be complete without mentioning some of the strange and bizarre human behavior it has triggered. In 1897, the House of Representatives in the State of Indiana unanimously passed a bill promulgating a new mathematical truth to the effect that Pi was 3.0. The State Senate in Indiana postponed the consideration of this measure indefinitely.

At the rise of Hitler in Germany, many Jewish professors were subjected by their "cultured" colleague to discrimination, vilification, discharge and eventual death in the extermination camps. In 1934 the eminent mathematician Edmund Landau was dismissed from his chair at Gottingen, after a virulent attack by the famous Prof. Ludwig Bieberbach on Landau's unique definition of Pi. Bieberbach wrote, "A people who have perceived how members of another race are working to impose ideas foreign to its own must refuse teachers of an alien culture" (Ref. III.1, see chapters 6 & 7). The great English mathematician G.H. Hardy issued an all-too-rare rebuke to Bieberbach for allowing his extreme racist views to color his scientific judgment.

There is a legion of clubs dedicated to the quasi-deification of Pi. The Harvard Math Club has a whole day dedicated to this number, complete with pie eating and throwing. The Albany Pi Club has several links including the Uselessness of Pi page and the Joy of Pi. There are a number of unofficial holidays to celebrate Pi. In America, Pi Day is observed on March 14 (3/14). Analysts, stock brokers and bloggers worry out loud about Pi Day. In European format, some Pi lovers dedicate "Pi Approximation Days," such as July 22 (22/7). Of the almost four million Google entries on the topic, two are outstanding with many relevant references, biographies and annotations (Ref.s III.2 & III.3).

2) Critique, Invective and Worse

Many of the comments on the "wrong" value of Pi in the Bible stem from either ignorance or an anti-religious Weltanschauung or a combination of both. In this section, a small sample of the more rational arguments is considered.

According to Prof. Steven Dutch of the Department of Natural and Applied Sciences, University of Wisconsin, "Now the Hebrews were not an especially technological society; when Solomon built his Temple he had to hire Phoenician artisans for the really technical work" (Ref. III.4). He proceeds to rip into Rybka (Ref. III.5) for using "fudge factors" and "finagle constants" to develop a formula resulting in a value of Pi=3.143. A most amazing statement follows:

> "The decimal expansion of Pi never ends and never repeats to infinity (this would have been a great place to put such a statement, which would have been utterly beyond the capabilities of the ancient Hebrews or even the translators of the King James Bible to have known. What a stunningly convincing proof of supernatural authorship it could have been!)."

I couldn't have said it better myself and look forward to his reaction upon reading this chapter.

Math textbooks have somewhat milder putdowns, as in Richardson's book: "For example, the Old Testament's estimate of 3 for the value of π may have been good enough for the roughly circular ox-cartwheels of that era" (Ref. III.6). In his paper, cited here as reference III.14, S. Belaga writes that he decided to publish his views on Pi after reading some dismissive comments in *The American Mathematical Monthly* by esteemed mathematicians who should know better (see Ref. III.7).

More vitriolic are the attacks from those entrenched in their tenured positions in academia. Consider the continuing saga of the Great Kansas Science Wars, in which professors of evolutionary biology, ecology, and other sciences find nothing better to do with their valuable time and experience than to take potshots at the Bible, such as "The Bible and the Flat Earth," "Noah's Ark Is A Wonderful Fable..." and, of course, "The Bible Misreports the Value of Pi." A group of Kansas University faculty even formed a club called FLAT to attack the Bible and Christianity. The "Pi error" was one of their favorite weapons (Ref. III.8).

In letters to a website called everything2.com (Ref. III.9), one particularly stands out. Entitled "How To Prove The Bible Is Wrong," this brilliant mathematician quotes Kings VII:23 and follows it with these original findings: "...you can see that this is a wrong description of Pi. It gives it as 3.0, which is WAY off for any decent calculation. I mean, the Bible is the word of God, right? Are you telling me God couldn't Figure out what Pi was?"

The fact that the ancient Hindus and Chinese used a value of

3 for Pi at about the same time the Hebrew Bible was being written has not elicited any similar ridicule and vitriol from these critics on the former peoples and their gods.

3) Apologia On Behalf of the Biblical Value

The many articles in defense of the "incorrect" value vary from a "common sense" approach to complicated and convoluted mathematical exercises. Natziger claims that the '10 cubits from brim to brim' refers to the outer diameter, whereas the 30 cubits refers to the inner circumference, i.e., the outer circumference was 31.4 cubits (using the thickness of the basin) (Ref. III.10). He concludes as follows: "Consequently, while this appears to be a straightforward error, a careful examination of the Biblical wording, along with some common sense, confirms that the Word of God is, in fact, infallible."

Hollenback (Ref. III.11) offers a similar interpretation of R.B.Y. Scott (Ref. III.12), discussed in detail and compared to the value in the Septuagint (LXX). A connection between the Alexandrian Jewish intellectuals and Archimedes is implied, but no conclusive resolution is achieved.

Morris uses the same argument (inner wall/outer wall) to achieve a value of $Pi = 3.1414$ and concludes with "God makes no mistakes, mathematical or otherwise" (Ref. III.13). He should have added "...but men generally do."

4) The Revealed Value of Pi in the Bible

The English translations of the Book of Kings (as cited in section 1) implied that the ancient Hebrews (and their God) did not know the correct value of Pi – certainly not with the accuracy possessed by their neighbors some 1000 years earlier.

Unfortunately, the translation of the original Hebrew into any other language (including the Greek of the Septuagint) results in loss of important and even crucial information. There are no digits in Hebrew; the letters themselves are numbers. Thus, every word written in Hebrew has a numerical value (*Gematria*).

The letters of the Hebrew alphabet were apparently used for numerical purposes well before the building of the First Temple of Solomon (see Ref. III.14).

The translations given in section 1 were based on the Hebrew sentences written as follows:

In Kings (1) VII:23:

ויעש את הים מוצק, עשר באמה משפתו עד שפתו **עגל** סביב וחמש באמה קומתו **וקוה** שלשים באמה יסד אתו סביב.

In Chronicles (2) IV:2:

ויעש את הים מוצק עשר באמה משפתו אל שפתו **עגול** סביב וחמש באמה קומתו **וקו** שלשים באמה יסב אתו סביב.

The following difference in the two versions is notable:

(a) In Kings, the word for *line* (קו) seems to be intentionally misspelled, as it is written with an extra letter (קוה). There are many instances in the Hebrew Bible where words are intentionally misspelled and the correct spelling is annotated on the margins or as a footnote.

The Jerusalem Bible (Koren Publishers Jerusalem) has such a correction on page 412.

(b) In Chronicles, the word for *line* is correctly spelled as קו.[6]

Shlomo Belaga (Ref. III.14) has pointed out that the rationale for the mathematical method used below and its historical origins have been lost to us.

The Rabbinical Exegesis (rediscovered by Max Munk, Ref. III.15) uses a ratio of the two spellings of *kav* (קו) to generate a value for Pi whose accuracy was only surpassed by Zu Chongzi in China 1,000 years later. Consider the following numbers:

Gematria of קוה (Kings) = 111

---

6. The word for *round* (עגול) is defective in the Kings version (no *vav*). This might be an additional clue which has not been investigated to date.

Gematria of קו (Chronicles) = 106

Take the ratio of 111 to 106 and multiply by 3 as follows:

$$\frac{111}{106} \times 3 = 3.141509 = \pi,$$

whose value is accurate to six significant Figures with an error of less than 0.0026%

(see Table 1).

In Table 1 this value is listed as the Jeremiah value, dated around 600 BCE. Secular authorities are in dispute concerning the dates of the composition of the Book of Kings.

Jepsen claims the date to be a few years after the destruction of the First Temple (around 580 BCE), whereas Fohrer argues for a date before 609 BCE, which would put it near the height of Jeremiah's prophecies. His Temple Sermon on the accession of King Jehoiakim is considered to have occurred around 609 BCE.

In fact, Jeremiah, together with his scribe, Baruch, is traditionally considered to be the author of the Book of Kings. Chronicles, on the other hand, was believed to have been composed several hundred years later by Ezra the Scribe, after the return from the Babylonian exile. Why was this "corrected" value of Pi omitted from Chronicles?

We can only speculate, since the real story is lost in the mists of our destroyed heritage. From Moses to Jeremiah, a span of 700 years, the knowledge imparted to Moses at Sinai was faithfully transmitted from one generation to the next. Much more than a disaster of physical dimensions resulted from the loss of Jerusalem and the Temple. By the time Ezra wrote Chronicles, much of the precious intellectual and spiritual heritage was lost. Our modern scholars (Refs. III.14 – III.16) are doing what they can to discover the basis for the exegesis, as described by Rabbi Max Munk in 1962. However, to date, no sources have been found (although some have claimed, without confirmation, that the current exegesis is from Rabbi Eliyahu of Vilna (the great *Vilna Gaon*).

Appendix III.1 – Huram of Tyre

"Even the floor was covered with gold" (1 Kings 6:30)

A metalworker called Huram from Tyre did the bronze work:

> King Solomon sent for a man named Huram, a craftsman living in the city of Tyre, who was skilled in bronze work. His father, who was no longer living, was from Tyre, and had also been a skilled bronze craftsman; his mother was from the tribe of Naphtali. Huram was an intelligent and experienced craftsman. He accepted King Solomon's invitation to be in charge of all the bronze work. (1 Kings 7:13–14)

Solomon wrote to King Hiram of Tyre:

> Now send me a man with skill in engraving, in working gold, silver, bronze, and iron, and in making blue, purple and red cloth. He will work with the craftsmen of Judah and Jerusalem whom my father David selected. (2 Chronicles 2:7)

King Hiram replied:

> I am sending you a wise and skillful master craftsman named Huram. His mother was a member of the tribe of Dan and his father was a native of Tyre. He knows how to make things out of gold, silver, bronze, iron, stone and wood. He can work with blue, purple, and red cloth, and with linen. He can do all sorts of engraving and can follow any design suggested to him. Let him work with your skilled workers and with those who worked for your father, King David. So now send us the wheat, barley, wine and olive oil that you promised. (2 Chronicles 2: 13–15)

[*Apparently, no one seems to troubled by the change from Naphtali to Dan as the origin of Huram's mother*]

References – Chapter III

III.1: Sanford L. Segal, *Mathematicians under the Nazis*, Princeton University Press, 2003

III.2: J.J. O'Connor and E.F. Robertson, "A Chronology of Pi," Sept. 2000 [http://www-history.mcs.st-and.ac.uk/HistTopics/Pi_chronology.html]

III.3: J.J. O'Connor and E.F. Robertson, "A History of Pi", August 2001 [http://en.wikipedia.org/wiki/History_of_pi#History]

III.4: Steven Dutch, "Pi in the Bible?", March 2002, [*www.uwgb.edu/dutchs/pseudosc/pibible.htm*]

III.5: Theodore Rybka, "Determination of the Hebrew Value Used for Pi", *Acts and Facts Bulletin of the Institute for Creation Research* (January, 1981)

III.6: M. Richardson, *Fundamentals of Mathematics*, MacMillan Co., 1941 (p.219)

III.7: E.T. Bell, *The Development of Mathematics*, McGraw Hill, 1945

III.8: Tom Willis, "The Continuing Saga of the Great Kansas Science Wars," Creation Science Association Newsletter for Nov./Dec. 1999 [http://www.csama.org/csanews/nws199911.pdf]

III.9: Letter from evans 927, "How to prove the Bible is wrong" [http://everything2.com/index.pl?node_id=701359]

III.10: Eric J. Natziger, "Biblical Value of Pi," Antioch Baptists Church, Knoxville, TN, 2003 [*www.learnthebible.org/molten_sea_value_of_pi.htm*]

III.11: G.M. Hollenback, "The Value of Pi and the Circumference of the 'Molten Sea' in 3 Kingdoms 7.10", Project Biblica, Vol. 85 (2004)

III.12: R.B.Y. Scott, "The Hebrew Cubit", *JBL* 77:209–210 (1958)

III.13: Dr. Henry M. Morris, "Pi and More in Torah", Defender's Bible, [*www.bibleprobe.com/pi.htm*]

III.14: Shlomo Edward G. Belaga, "On the Rabbinical Exegesis of an Enhanced Biblical Value of π," C.N.R.SAM1980. (See also: *www.math.ubc.ca/people/faculty/israel/bpi/bpi.html*).

III.15: Rabbi Max Munk, "Three geometry problems in Tanach

and Talmud" (Hebrew), *SINAI*, 51: 218–227 (Harav Kook Institution, 5722).

III.16: H. Roiter, "La mer d'airain du Roi Salomon et le nombre π (Pi)", *Kountrass* 7, no. 18:10 (1993)

Chapter IV
Chemistry and the Bible

This chapter has three parts, each dealing with a different aspect of chemistry in the Bible. The first section starts with instructions given Moses at Mt. Sinai concerning the wearing of fringes at the corners of garments. These fringes are to be dyed with the color of sky-blue (tekhelet). The commentaries attempt to resolve some of the apparent contradictions in the quoted passage. The source of tekhelet and its relation to Tyrian Purple is described from an historical and scientific viewpoint.

Moses' blessing to the tribe of Zevulun provides a clue to a possible source of this rare sky-blue (and other treasures like glass) to be found in the beaches and sea coast of northern Israel. This first section concludes with a discussion of the flag of Israel, whose colors and stripes are said to be based on the sky-blue dye.

The next section deals with the structure of water and its anomalous properties. The possible prediction of the molecular structure of water by the written Hebrew word for water, mayim, is presented as one of the secrets of science embedded in the Bible. The chapter concludes with a brief comparison of the insulating properties of snow and fleece as described in the Book of Psalms.

Hidden Light: Science Secrets of the Bible

A. BIBLICAL BLUE (TEKHELET): ORIGINS AND CHEMISTRY

1) Questions on the Book of Numbers xv:38–39

In the fourth book of the Pentateuch, (the Book of Numbers), God tells Moses that all the males of Israel should observe the *mitzvah of Tsitsit* (ציצית – the commandment to wear fringes, or tassels):

> "דבר אל בני ישראל ואמרת אלהם ועשו להם ציצת על־כנפי בגדיהם לדרתם ונתנו על־ציצת הכנף פתיל תכלת. והיה לכם לציצת וראיתם אתו וזכרתם את כל מצות ה' ועשיתם אתם ולא תתורו אחרי לבבכם ואחרי עיניכם..."

> "Speak to the children of Israel, and tell them to make for themselves tassels (or fringes) on the corners of their garments throughout their generations, and attach (or affix) upon the fringe of each corner of your garments a thread of sky-blue (*p'til tekhelet*) and it shall be to you as fringes, that you may look upon *it*, and remember all the commandments of the Lord, and do them; and that you seek not to go astray, [touring] after the inclinations of your heart and your eyes..."

There are several questions on this portion, which comprises the third part of the *Shema* prayer (said twice daily), proclaiming the Oneness and Unity of God.

    a. Do the words *tekhelet* or *p'til tekhelet* (sky-blue thread) appear anywhere else in the Bible?

    b. Why is this placed here (at the end of *Parashat Shelach*) with no apparent connection to any of the preceding material?

    c. The word ציצית (*tzizit*), meaning fringes or tassels, is in the feminine. So why is its objective form, "it" – אתו (*otoh*) written in the masculine singular (*it*)?

    d. Why is the word ציצית spelled incompletely as ציצת (missing a *yod*)?

    e. How can one remember and keep all of God's commandments just by "seeing" these tassels?

2) Commentaries on the Questions

a. There are a number of other places in the Bible where the term *tekhelet* appears. The descriptions of the *Mishkan* (Desert Tabernacle) and of the clothes of the High Priest were given by God to Moses, as follows:

I. *Exodus XXVI:1*

"ואת המשכן תעשה עשר יריעת שש משזר ותכלת וארגמן ותולעת שני..."

"Thou shalt make the tabernacle with ten curtains of fine twined linen, and sky-blue, royal purple and crimson…"

II. *Exodus XXVI: 31*

"ועשית פרכת תכלת וארגמן ותולעת שני ושש משזר מעשה חשב..."

"And thou shalt make a partition veil of sky-blue, and royal purple, and crimson and fine twined linen…"

III. *Exodus XXVIII:5*

"והם יקחו את הזהב ואת התכלת ואת הארגמן ואת־תולעת השני ואת השש"

"And they [the skilled artisans] shall take gold, and sky-blue, and royal purple, and crimson, and fine linen".

The Torah then goes on to describe how these materials were used to create the *ephod* (a sort of cape or vest), the breastplate, the robe with its pomegranates and bells and the forehead-plate to be worn by Aaron, the High Priest.

IV. *Exodus: XXVIII:36–37*

"וְעָשִׂיתָ צִּיץ זָהָב טָהוֹר וּפִתַּחְתָּ עָלָיו פִּתּוּחֵי חֹתָם קֹדֶשׁ לַה׳. וְשַׂמְתָּ אֹתוֹ עַל־פְּתִיל תְּכֵלֶת וְהָיָה עַל־הַמִּצְנָפֶת...."

> "...make a plate of pure gold, and engrave upon it, in the same manner as a signet ring 'Holiness to the Lord.' Then attach to it a twist of sky blue wool so that it can be upon the headdress..."

One could reasonably ask why the Torah (and this author) spend so much time and space on these detailed descriptions, especially seen in the constant repetition of the terms sky-blue, royal purple and crimson. Whenever the Bible uses such repetitive language, it is usually interpreted as the key to a hidden message. In this case, the chemical compositions of *tekhelet* (sky-blue) and *argaman* (royal purple) may turn out to be one of the Secrets of Science found in the Bible. There may also be a connection to the color wheel discussed in section 5 on the Physics of Color.

b. We ask if there is any connection between the commandment to wear *tzitzit* with *p'til tekhelet* and the rest of *Parashat Shelach*, most of which is concerned with the episode of the twelve spies, ten of whom came back with an evil report on the Land of Israel, which doomed the nation to forty years of wandering in the desert. The connection, at least superficially, is in the last sentence quoted, *v'lo taturu* – don't go astray ('touring'). At the start of *Shelach*, Moses instructs the spies: *la'tour et ha'aretz* – to spy out or explore the land. The root (*tour*) is actually repeated there three times, as if to flag the same word appearing much later at the end of the same *parasha*.

c-e. In a critique of Rashi on question e. (how can one remember to keep all of God's commandments just by seeing these tassels), Rambam also provided answers to questions c. and d. Rashi invoked *Gematria* (numerical value of letters) by pointing out that the numerical value of tzizit (ציצית) is 600. According to a tradition now widely accepted by the majority, there are eight threads and five knots on each tassel so that the total equals 613, which is the sum total of all the positive and negative commandments derived from the Torah. Rambam, however, points out that ציצת is spelled defectively (missing a *yod*,) so that its numerical value is 590, not 600. Also, there had been controversy on the number of knots and threads per tassel. In his Mishneh Torah (The Book of Adoration), Volume II, Chapter II, Rambam writes, "... *tekhelet* (blue thread) whenever it is mentioned in the Torah is wool dyed azure, the color of the firmament seen in a blue sky."

Rabbi Meir in the Gemara of Menachot 43b had written "Why is tekhelet so different from all other dyes? For tekhelet is like the the sea and the sea is like the sky and the sky is like the Heavenly Throne."

Therefore, it is the sight of this Biblical Blue twisted thread (*p'til tekhelet*) which reminds a person of the sea, the sky (heaven) and the Throne of Glory, resolving the grammatical problem of question d., since p'til is masculine singular. The concept of the Throne of Glory stems from a highly metaphorical phrase in Exodus XXIV:9–10:

"ויעל משה ואהרן נדב ואביהוא ושבעים מזקני ישראל ויראו את אלהי ישראל ותחת רגליו כמעשה לבנת הספיר וכעצם השמים לטהור"

> "Moses went up along with Aaron, Nadav and Avihu and seventy of the elders of Israel. And they saw a vision of the God of Israel and under his "feet" was something like sapphire brickwork like the essence of heaven for purity"

This verse has elicited much commentary and numerous attempts to rationalize the anthropomorphic connotations. Some commentaries state that they saw the base (feet) of the Throne of Glory (also known as The Throne of God). There are references ranging from the prophets (Isaiah VI, Ezekiel I and Daniel VII) to kabalistic, Talmudic and Midrashic sources. Some identify the color of clear sapphire as connected to sky-blue tekhelet. A website for Gemnation states flatly, "Sapphire and the color blue tend to be synonymous" (*www.gemnation.com/base?processor=getPage&page*Name=sapphire_color).

3) Biblical Blue (tekhelet) and Tyrian Purple (argaman) – History and Sources

The two dyes which were used in ancient times by royalty and nobles were Tyrian (or Royal) Purple and Sky-Blue (*tekhelet*), both extracted from mollusks found along the Eastern Mediterranean coast, from Tel Dor (in northern Israel) to Tyre in Lebanon. Tyre and Sidon became stopovers on the Silk Route from the Far East as a result of the dye works there. Archaeologists have found great quantities of shells along this coast, attesting to a flourishing industry which seemed to vanish a few years after the Arab Conquest in the seventh century.

Scholars cite a number of factors leading to this decline, beginning with the Roman occupation and the Byzantine Christian persecutions and massacres in 628, about a decade before the arrival of the Muslim armies. Many Jews fled Israel, and this sudden loss of skilled artisans may have contributed to the collapse of the country's industry.

There is voluminous Talmudic literature (e.g. Megila 6, 42 and 43, Menachot 44, Shabbat 26, Sanhedrin 12) and a number of Midrashim, such as Midrash Rabba, which identifies the source of the tekhelet as *chilazon*, a sea creature. The various descriptions were not always consistent, which has led to considerable dispute and controversy to the present day. There is no explicit mention of *chilazon* in the Bible as the source of the *tekhelet*. However, the Talmud in Megila 6a cites verses Deuteronomy XXXIII:18–19 as representing the *chilazon*. Moses there is addressing each of the tribal divisions on the final day of his life and has already given his blessings to Reuven, Levi, Judah, Benjamin and Joseph. It is now the turn of Zevulun and Issachar:

ולזבולן אמר שמח זבולן בצתך ויששכר באהליך
עמים הר יקראו שם יזבחו זבחי צדק
כי שפע ימים יינקו ושפני טמוני חול

"And to Zevulun he said; rejoice Zevulun in your expeditions and Issachar in your tents. They shall summon nations to the mountain and there they (the nations) will offer righteous sacrifices – they (Zevulun) will be nourished by the abundance of the seas and *by the hidden secret treasures of the sands*."

The Talmud (Megila 6a) says that these (italicized) words refer to the *chilazon* as the source of the costly *tekhelet*, a view supported by Rashi in his commentary on this sentence. There is a difference of opinion whether the *chilazon* was to be found on the sands of the beach, the sands of the mountains or in the sandy seabed near the coast.

The Talmud and Rashi point out that the source of the sky-blue dye was not the only secret treasure to be extracted from the seashore; for example, white glass was another precious inheritance for Zevulun, to make up for the lack of fields and orchards in his territory. Targum Yonatan (translation of the Pentateuch into Aramaic), written around the middle of the first century, states that Moses' blessing to Zevulun refers to glass. Others (Chizkuni) cite the riches to be found in ancient shipwrecks buried in the sand which would be exposed at low tide. When Rashi wrote his commentary, he had the benefit of 12th century knowledge of glass making and glass-blowing, whereas Moses' blessing and prophecy are being bestowed upon Zevulun 2500 years earlier.

In secular history books, there is little or no mention of Zevulun as a sea-faring people, whereas the Phoenicians are recognized as the sailors of antiquity. According to some historians, the Phoenicians originally migrated from the Persian Gulf around 3000 BCE and settled on the northern coast of Canaan. The earliest archaeological traces of their presence in the Canaanite coastal areas date from late 13th century BCE.

According to Pliny (in his "Natural History"), the Phoenician sailors accidentally discovered the art of glass manufacture. Upon landing at a beach near the mouth of the River Belus in northern Israel, they lit a big bonfire on the sand and made preparations for their evening meal. Since they were unable to find rocks to support their pots, some sailors went back to the ship and brought back some saltpeter cakes (nitrum in Pliny).

When the fire had died down, they were surprised to find shiny glass globules in the ashes. Glass is an amorphous solid formed when the temperature of a supercooled liquid drops. In this story, the heat of the fire caused a chemical reaction between the beach sand, which is almost pure silicon dioxide (SiO_2), and the saltpeter (potassium nitrate-KNO_3), to form silicates.

An alternative version opines that the sailors were traders in natron bricks (sodium carbonate) which were used to support their pots. A third point of view states that the reactants were sand and the alkali hydroxides exhuded by the burning wood ($NaOH$ and KOH). Most authorities claim that Pliny's account is apocryphal, since glass was fabricated in Egypt and Mesopatamia as early as 2500 BCE, well before the Phoenicians sailed the Seven Seas.

It is also doubtful whether the heat of a beach fire could produce temperatures high enough (2000 degrees) to fuse the sand with the cake chemicals to form glass. However, it is not the fire which initiates the chemical reaction, but the hot glowing coals, which are in direct contact with the sand. Some skeptics also claim that lime (calcium carbonate) was a necessary ingredient in the manufacture of glass. Others conjecture that seashells (maybe even those of the *chilazon*) could have provided the calcium. Modern attempts to reproduce this discovery have yielded inconclusive results.

According to the prophetic blessings of Moses and Rashi's comments on "white glass," one might posit that it was the sailors of Zevulun who discovered the art of glass manufacture. Pliny was a great admirer of the Phoenicians, whereas his feelings about the Jewish inhabitants of the region may have been colored by his good friendship with Vespasian and Titus. Some unsubstantiated accounts even have him present at the siege and destruction of Jerusalem (nine years before his death at the eruption of Vesusius in 79, as described in Chapter 11).

It was not unusual for the Greek and Roman historians to ignore or even denigrate the contributions of Israel. According to the Book of Joshua, Zevulun occupied their territory sometime at the beginning of the thirteenth century. Some maps show the territory of Zevulun extending along the coast from Dor northward along the coast to Asher, whereas others show his territory only reaching the coast around present-day's Nahariya, and running in a narrow coastal strip parallel to the lands of Asher and Naphtali up to Tyre and Zidon (see Aryeh Kaplan's The Living Torah, Plate 12, page 1040, with the caption "Portions of the Tribes" below). The accounts in Genesis, Joshua and Judges are not clearly in agreement on this issue.

In Genesis XLIX:13, when Jacob gives his final blessings to his 12 sons, the progenitors of the 12 Tribes of Israel, he says to Zevulun,

"Zevulun shall settle by the seashore. He shall be at the ships harbor and his last border shall reach Zidon." In Joshua XIX:10–11 it states "…and the border of their inheritance was unto Sarid. And their border went up westward (la-yama)," which some interpret as a seaward direction, reaching the Mediterranean at Akko. However, in Judges V:17–18 in the Song of Deborah following the victory of Barak over Sisera, we read, "Asher dwells by the sea" whereas Zevulun is only mentioned for his bravery and valor in battle. The friendly relations between the Phoenicians and these northern tribes are well documented in the Bible, culminating in the cooperation between Solomon and Hiram in the construction of the First Temple. According to the Encyclopedia Judaica (Volume 16: p. 951), a significant portion of Zevulun was not deported during the Assyrian conquest of the Ten Tribes in 720 BCE. It may have constituted the major part of the Jewish population in the Galilee up to the wars with Rome. There is also a tradition that some of those exiled by the Assyrians may have retuned to their ancestral lands under the guidance of Jeremiah.

Whether or not Pliny was describing the sailors on that remote beach as Phoenicians when he meant Israelites, there remains a gap of more than 1000 years between the manufacture of glass in Egypt and Mesopotamia and its purported discovery on the coast of Israel. Rashi and the Talmud, in their thoughts on Zevulun's inheritance, may have provided a solution to this contradiction.

There is a lengthy discourse in the Book of Job (XXVII:12–17) in which Wisdom is considered more precious than gems, and which concludes with "more precious than gold or glass." This phrase gives us the insight that when Job was written, glass was a treasure on a par with gold.

Most secular authorities are of the opinion that the Book of Job was redacted sometime around the ninth century BCE. This would mean that glass had been discovered in Israel when Zevulun's ship explorations were at their peak. The River Belus in Pliny's account has been identified as present day River Naaman, which empties into the Mediterranean Sea at Acre, where the sand is of exceptionally fine quality. In fact, there is a large glass factory there today which exports glass all over the world (Phoenicia Glass Works).

In his Wars of the Jews, Book II, x.2, Josephus waxes rhapsodic on the fine sand at this location, interrupting his account of the Roman general Publius Petronius. Gaius Caesar (Caligula) had ordered Petronius to march out of Syria with two legions and Syrian auxiliaries to teach the rebellious Jews a lesson (about 41 CE – 25 years before the Great Revolt). Caligula, a self-proclaimed deity, had issued edicts throughout the Roman Empire that his subjects were to worship his statue. All obeyed except the obstinate Jews. Petronius reached Ptolemais (modern day Acre) at the mouth of the River Belus, which in the words of Josephus, "…hath near it a place no larger than 100 cubits (150 feet) which deserves admiration; for the place is round and hollow and affords such sand as glass is made of."

In a related footnote, Josephus adds about this place: "…. whence came that sand out of which the ancients made their glass is a known thing in history, particularly in Tacitus and Strato and more largely in Pliny."

As a contemporary of Pliny, Josephus would not have known what we know today about the discovery of glass in Mesopotamia and Egypt 2500 years earlier. It is highly probable that the details of glass manufacture in Mesopotamia were a closely held trade secret and that the "accidental discovery" described by Pliny may have broken its monopoly in the ancient world.

A great delegation of Jewish notables, along with their

wives and children, implored Petronius to explain to his emperor that the whole country would go up in flames rather than have a graven image placed in the Temple. Petronius was so moved by the piety and entreaties of the Jewish people that instead of erecting a statue in the Temple, he led his troops back to Antioch. The outraged Caligula dispatched a letter to his errant Governor of Syria, ordering him to commit suicide. The ship carrying this letter was delayed by storms, whereas a later letter arrived several weeks before the original decree, containing the news that Caligula had been assasinated!

In addition to the hidden treasures of the *chilazon*, shipwrecks, and glass hinted at in the blessings of Moses, modern science has produced optical glass fibers, silicon photodiodes and transistor electronics from these same sands. My teacher and rabbi Daniel Lapin once commented that those of us who worked on the early designs of fiber optic transmission systems may have one of our ancient ancestors from Zevulun.

Mention of the *chilazon* disappears from all texts after the last of the Amoraim (around the year 500) and is lamented by The Midrash Tanchuma Shelach (ca. 750). As a result of intensive research and investigations by zoologists, chemists, archaeologists and historians, there is now some general idea that the source of the *tekhelet* may have been the *Murex Trunculus*, one of three species of murex found in the eastern Mediterranean.

An organization called *P'til Tekhelet* was formed in an effort to provide *tekhelet* to the general public (Refs. IV.1 and IV.2). Starting in a kitchen in the town of Efrat, it now has a factory at the settlement of Kfar Adumim, just a 20-minute drive from Jerusalem. Since murex is a protected species in Israeli waters, it imports the murex trunculus from Spain and Greece and processes the secretions to provide *p'til tekhelet* on the fringes of prayer shawls and tzizit undergarments. Although the findings and claims of the P'til Tekhelet Foundation are highly suggestive, their conclusions have not been universally accepted (Ref. IV.3). According to some modern researchers, it takes 8,000 murex sea snails to produce one milliliter of dye (Ref. IV.4). In this same reference, the author identifies the Murex Brandaris as the primary source of the Tyrian Purple. In fact, it was a French zoologist, Henri de Lacaze-Duthiers, who determined in 1858 that three mollusks in the Murex family were capable of producing the dyes which commanded such a high price in antiquity (which Pliny, in his Natural History, described as a "mad lust for purple").

4) The Chemical Structures

In 1887, the Rabbi of Radzin, Poland (Rabbi Gershon Chanoch Leiner), began a year-long journey visiting many ports in the Mediterranean as part of his obsession with finding the source of the long-lost *tekhelet*. He was the author of a number of learned treatises on the subject, one of which had the evocative title, "Sfunay T'munay Chol," referring to the buried treasures to be found on Zevulun's seashore.

In an aquarium in Naples he spotted the cuttlefish (Sepia Officinalis) secreting a dark fluid. The Rabbi had read the Rambam's description that the fluid secreted by the *chilazon* was dark at origin and only later turned blue. Thus, when he spotted the cuttlefish emitting this black substance, he was convinced he had found the source of tekhelet. Another factor in the Rabbi's identification of the cuttlefish as the long-lost *chilazon* is the Talmud in *Avodah Zara* 28b, which states that the *chilazon* was one of the treatments for hemorrhoids. In the nineteenth century cuttlefish ink (sepia) was sold as a cure for this malady and is available even today in tablet form. After consultations with some Italian chemists, he was able to transform the squid's black secretions to a beautiful sky-blue. Within a year, 10,000 of his followers were wearing blue fringes and a dye factory was in full operation at Radzin. The recipe for the transformation involved heating a squid's "ink" to a high temperature and then ADDing iron filings to the hot liquid. About 30 years later it was determined by Dr. Isaac Herzog during his thesis research that a chemical reaction between the iron (Fe) and the organic molecules of the squid ink produced ferric ferrocyanide, the famous Prussian Blue (Dr. Herzog later became the Chief Rabbi of the State of Israel).

The chemical formula of Prussian Blue is:

$Fe[Fe^{3+} + Fe^{2+} + (CN)_6]_3 = Fe_7(CN)_{18}$

Therefore, any organic compound containing nitrogen may have given the same result. The structure of Prussian blue was only elucidated in the last decades of the 20th century (Ref. IV.5).

Although the factory at Radzin was destroyed during World War II, the followers of the Radziner Rebbe made it to Israel where, thanks to the correspondence records of Chief Rabbi Herzog, they were able to establish a new dye plant. Under no circumstances did Dr. Herzog or anyone else ever suggest that Rabbi Leiner had knowingly misled his followers and the halachic world. In fact he was misled, first by the fact that Rambam utilized the words *dag* (fish) and *shachor* (dark or black) in his description of the source and color of the *chilazon* secretions, and then by the well-intentioned advice of the chemists. Incidentally, despite its name, the cuttlefish is not a fish, but a mollusk of the order Sepiida (which also includes squid, octopus and nautilus). Despite his mistake, we need to recognize that it was the Rabbi's dedicated searching which led to a revival of interest in a subject which had lain dormant for almost 1000 years.

During the same period as the Radziner Rebbe, a number of chemists were investigating the structure of the indigo molecule and its derivatives. Indigo originally was extracted from a plant called the *Kali Ilan*, which was shunned by rabbinical authorities as a source of "False *Tekhelet*." The Talmud describes chemical tests which were used to distinguish the real tekhelet from the much cheaper plant derivative. To date, attempts to reproduce these tests have produced inconclusive results.

The work of Friedlander and his colleagues in 1912 (Ref. IV.6) laid much of the basis for the later researches, such as those of Formanek (Ref. IV.7) and the more recent discoveries of Cooksey, et al. Friedlander was able to extract 1.4 grams of Tyrian Purple from 12,000 sea snails (the murex brandaris). In a classic review article, Cooksey states unequivocally that 6, 6'di-bromo indigo (DBI) is a major component of the historic pigment Tyrian Purple, also known as Royal Purple, and clearly offers a word of caution:

Hidden Light: Science Secrets of the Bible

"Surviving details of the ancient process are insufficient to explain the chemistry involved" (Ref. IV.8).

According to these findings and earlier work, DBI does not exist in the mollusk – instead it is generated by precursors (chromogens) contained in the hypo-branchial gland of the murex. The structure of DBI is shown in Fig. IV.1 (top).

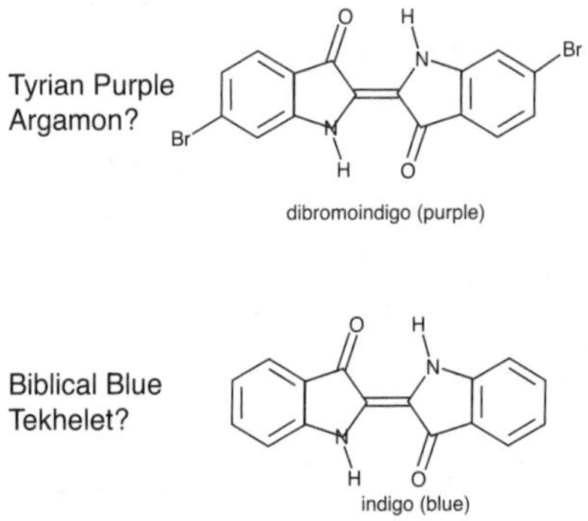

Tyrian Purple Argamon?

dibromoindigo (purple)

Biblical Blue Tekhelet?

indigo (blue)

Fig IV.1 – Showing the difference between Tyrian Purple and Biblical Blue

There seems to be some confusion in the literature on the various secretions obtained from murex brandaris and murex trunculus. The brandaris mollusk produces almost pure Tyrian Purple whereas trunculus secretes a mixture of DBI and indigo. In order to obtain unadulterated sky blue, a combination of exposure to sunlight with chemical processing is employed to remove the DBI from the mixture extracted from the trunculus. According to Ref. IV.9, brandaris secretes 83% DBI and zero percent indigo, whereas trunculus produces a mixture comprising 55% indigo, 35% DBI plus some transition forms such as monobromindigo (see Fig. IV.1). Whereas the indigo from the plant Kali Ilan is extracted in pure form, it is possible that processing from the *murex trunculus* leaves a residue of DBI which inhibits chemical reactions.

For those not conversant in organic chemistry, a brief explanation is in order:

Each of the hexagonal rings represents a benzene molecule whose structure is shown in Fig. IV 2. Unless otherwise noted, each carbon atom has a single hydrogen attached. To simplify the complicated structural diagrams of organic molecules (such as DBI in Fig IV.1), the carbon atoms together with their associated hydrogens are not explicitly shown (Benzene was discovered in 1825 by Michael Faroday, who isolated it from coal gas).

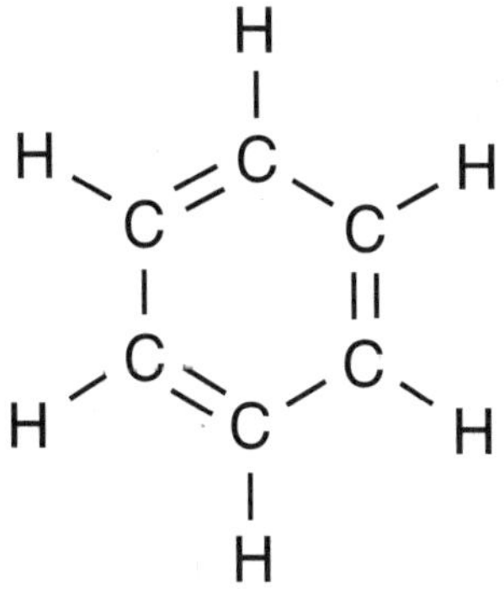

Fig IV.2 – The Structure of benzene

The formula of benzene (C_6H_6) baffled scientists for almost fifty years, since no explanation was found to account for all the valence bonds (carbon has valence of four and hydrogen has valence of one). One of my earliest memories as a young chemist (at age 15) was reading a text written before World War I, which described how Friedrich August Kekule von Stradonitz was able to deduce the structure after struggling with the problem for a number of years. Kekule had spent many years in the study of carbon bonding and was dozing one night in front of his fireplace. As he observed the smoke wisps through his droopy eyelids, he imagined that he saw snakes curling around and eating their own tails (there is in fact an ancient symbol depicting a serpent swallowing its own tail, called the Ouroboros).

He awoke with the realization that the benzene structure was a cyclic ring, as shown in Fig. IV.2, one of the most seminal discoveries in the history of chemistry.

5) The Physics of Color and Photochromic Processes

The color of the dyes extracted from the murex family depends on a number of variables. Very little work has been done on the substances themselves. Most of the attention has focused on the optical properties of DBI and indigo, either in solutions with various organic solvents or in their interactions with wool and silk. One of the most interesting ancient records cites the observation that wool dyed with the murex secretions on a cloudy day is purple, whereas the dye process on a bright, sunny day produces a sky-blue color (see also Spanier, et.al, Ref. IV.9). Evidently, the purple color is the result of dyeing with the DBI component, whereas exposure to sunlight with a strong ultra-violet content causes a photo-chemical reaction in which the bromine atoms are stripped away, converting the DBI to indigo (see Fig. IV.1). This is called a photochromic reaction, where light of sufficient photon energy and intensity produces a change in color due to a change in the chemical composition.

Generally, we speak of color as seen by the eye under the normal illumination of our environment, whether by sunlight or indoor artificial lighting (white light). What our eyes perceive as the color of an object is the result of several processes which occur when "white light," containing many wavelengths between 400 and 800 nanometers, is incident on the object.

Certain wavelengths are transmitted through the materials and may even be partially reflected by the "impedance mismatch" (to use a term from electronics). These wavelengths may also undergo multiple scatterings. Others may be totally absorbed, leaving a predominant component of the reflected beam, which impinges on the human eye and produces the sensation of color. The details depend on the index of refraction of the colored object and its variation with wavelength in the visible range. In the visible light range there is negligible absorption by indigo until an absorption peak appears at plus 600 *nm*, corresponding to red and orange. The residual reflected and transmitted beams give the substance its characteristic sky-blue color.

The variation of the absorption spectrum of DBI and indigo

with solvent, concentration, temperature, etc., has been extensively investigated since the pioneering work of Friedlander and Formanek. The effect of solvent on the absorption maximum is shown in Table IV.1 (below).

The P'til Tekhelet Institute has widely publicized the recent results of two Belgian scientists (J. Wouters and A. Verhecken) (Ref. IV.10), whose absorption spectrum results are shown in Fig. IV.3.

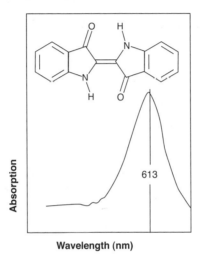

Fig IV.3 – Absorption spectrum of tekhelet dye (Ref. IV.10)

Dr. Baruch Sterman in Ref. IV.2 points out that the absorption peak at 613 *nm* is more than sheer coincidence.

Recapitulating the questions of Section 1 on the *tzitzit* Commandment:

(b) If ציצית is feminine plural, why is the objective form written as mascular singular?

(c) Why is the word ציצית spelled defectively (the second *yod* is missing)?

(d) How can one remember and keep all (613) of God's commandments just by seeing these fringes?

Sterman and his colleagues posit that the Torah is telling us that *it* (*otoh*, אתו) is the sight of the sky-blue color (the *P'til Tekhelet*) which reminds us of God's commandments and not the sight of the feminine plural *tzitzit*. This, together with the defective spelling of *tzitzit*, would appear to favor Rambam's interpretation over Rashi. The "coincidence" of the absorption peak corresponding to the derived 613 commandments is considered ADDitional confirmation that the murex trunculus is indeed the *chilazon*.

There are a number of problems with these basic conclusions:

(1) Several combinations of DBI and related compounds in various solvents also show absorption peaks at 613 *nm* or close to it from the earliest investigations (see Table IV.1).

(2) It has been noted that the meter (and thus the nanometer) is a purely arbitrary unit of length with no halachic significance and that wavelength expressed in other units (milicubits, microinches or Angstroms) would not be 613. On the other hand, the defenders of Sterman's interpretation say this result shows that the metric system, as introduced by the French revolution, is somehow a preferred system of measurement in halachic matters. The meter was originally defined as one-ten-millionth of the distance from the North Pole to the Equator, so it is indeed based on the dimensions of our very own Privileged Planet. Also, some argue that if the meter is a man-made construct, so is *Gematria*, on whose basis Rashi derived the 613 commandments.

Although the evidence cited to date is suggestive, many authorities do not accept the hypothesis that the murex trunculus is indeed the *chilazon* of antiquity nor that indigo is tekhelet (see Ref. IV.4).

Table IV:1 – Absorption spectrum maxima as a function of bromination and solvent

Bromine substituents	Solvent	λ max. (in *nm*)	Reference
6, 6'	tce*	585	6
6, 6'	xyl**	590.5	7
None (indigo)	tce	605	8
4, 4'-	tce	613	6
None (indigo)	pyr	613	10

* tce = trichloroethylene
** xyl = xylol

6) Tekhelet and the Flag of Israel

The 1948 Proclamation for the Flag of Israel states that the two wide stripes are to be dark blue (*Tekhelet Ke'he*) and the *Magen David* (Star of David) needs to be sky-blue (*Tekhelet*) all on a white background. This combination of blue and white is reminiscent of the commandment on *tzizit* discussed in Section 1 and thereafter. The *talit* (regular prayer shawl) is worn only during prayer, whereas the *talit katan* (small talit, or tzizit) is worn as an article of clothing all day, every day, and whose fringes serve as the fulfillment of the commandment. There are no stripes on the *talit katan* whereas the *talit* worn during prayer usually has two stripes (blue, black or dull white) on a white shawl background. Some sources trace the black stripes to the same misreading of Maimonides which led the Radziner Rabbi to identify the cuttlefish as the *chilazon*.

The stripes on the state flag were a dark navy blue until the 1960's when they were changed to a lighter shade of blue. The connection between *talit* and flag was vividly depicted by Ben Hecht in his play, "A Flag is Born," written at the end of World War II.

An elderly Jewish couple who barely survived the Holocaust give the old man's *talit* to a Jewish youth just before they die. The young man accepts the gift and converts it to a flag, as he sets off to Fight for the remnant of his people in the nascent State of Israel.

Yasser Arafat and those of his persuasion (including some anti-Zionist Jews) claimed that the two blue stripes on the Flag of Israel "secretly" represented the Nile and Euphrates Rivers, as a symbol of Israel's expansionist ambitions. These secret ambitions were said to be based on Genesis XVII: 18 where God promises Abraham: "…unto thy seed have I given this land from the River of Egypt to the Great River, the River Euphrates." The promise was actually fulfilled during the reigns of David and Solomon.

B. THE STRUCTURE OF WATER AND ITS AMAZING PROPERTIES

1) The Water Molecule

Water has a number of unique and anomalous properties which are essential to life as we know it on our privileged planet. It is the simplest compound formed by the two most reactive elements in the Universe, hydrogen (H) and oxygen (O), which combine to form water according to the following equation: $2 H_2 + O_2 \leftrightarrow 2H_2O$.

Two molecules of hydrogen plus one molecule of oxygen react to produce two molecules of water.

The structure of the water molecule depends on the valences of hydrogen and oxygen, which in turn are determined by the atomic number of each element. The atomic number is the number of protons in the nucleus equal to the total number of electrons in orbit around the nucleus. In an electrically neutral atom, the number of electrons is equal to the number of protons. Hydrogen has atomic number one whereas oxygen is atomic number eight. Before the advent of quantum mechanics, the Bohr-Rutherford model was used to explain the chemical valence of each element by the empirical rule that the electrons orbiting the nucleus were arranged in shells of 2, 8, 18…Quantum mechanics provided a theoretical basis for the shell structure, combining solutions of the

Schrodinger Wave Equation with the Pauli Exclusion Principle (see Appendix 1.4 to Chapter 1c).

Hydrogen with its solitary electron has a valence of one whereas the oxygen atom with six electrons in its outer shell has a valence of two. A covalent bond is formed when two atoms share a pair of electrons. In the water molecule, the single electron of each hydrogen is shared with one of the six outer shell electrons of oxygen The mutual electrostatic repulsive forces between the four remaining electrons cause the water molecule to assume a quasi-tetrahedral structure, where the bond angle is 104.5 degrees.

The electrons are concentrated around the oxygen nucleus, leaving each of the hydrogen atoms partially positive. The water molecule is electrically neutral but the non-uniform distribution of charges constitutes an electric dipole, which leads to the phenomenon of hydrogen bonding between adjacent water molecules. The partially negative oxygen on one molecule is weakly attracted to the partially positive hydrogen on its nearest neighbor. The energy of such dipole-dipole linkages is significantly less than the bonds in the water molecule itself. This so-called "hydrogen bonding" energy is about the same as the thermal energy at room temperature (approximately $\frac{1}{40}$ of an electron-volt).

The linked aggregates which result from hydrogen bonding are called clusters, containing up to 100 molecules near the freezing point and less as the temperature increases. At room temperature, these linked aggregates are continually breaking apart and reforming. Most chemists believe that it is the clustered structure of water which is responsible for its anomalous properties, such as its high boiling point, its high heat capacity, its thermal conductivity and the unique change in density at 4 degrees Centigrade.

Ice floats – there are very few substances where the solid form has a lower density than the liquid form and none except water in the natural state. Liquid water has been visualized as a seething mass of water molecules in which hydrogen-bonded clusters are continually forming, breaking apart and reforming.

There are perhaps 20 models in the science of water seeking to explain its remarkable properties and much controversy on any attempt at a general explanation. An eminent immunologist who did some of his early research on the properties of water once confided to me that there were two areas which he would never again revisit, the role of calcium in cells and the structure of water.

One of the most controversial properties of water is its memory for solutes which have been diluted down to infinitesimal concentrations. Homeopathic physicians and others claim that clusters formed around solute ions may somehow be responsible for this memory (Ref. IV.11).

2) Water in the Bible

The Hebrew word for water (*mayim* – מים) appears in the second sentence of the Bible "...and the Divine Presence hovered over the surface of the waters." The word is written in this plural form in all the 180 times that it occurs in the Bible, even when it refers to water in its singular sense. Perhaps this is a hint at the clustering nature of the liquid substance, which gives water its unique properties.

About 25 years ago, my teacher, Rabbi Daniel Lapin, pointed out an interesting "coincidence" between the Hebrew word for water and the geometry of the water molecule (which he had learned from his father, Rabbi Avraham Chaim Lapin, z.l.). The spelling of *mayim* in both biblical and modern Hebrew and its comparison with the H-O-H structure is illustrated in Figure IV.4. Consider two imaginary lines connecting the centers of each outer letter *mem* (מ) to the center of the *yod* (י). These connecting lines form an angle

Fig IV.4 – The word מים (mayim) predicts the bond angle of the water molecule

close to 105 degrees. If the *mems* represent hydrogen atoms and the *yod* is oxygen, the ancient word *mayim* contains within its written form the secret of the water molecule's structure, only recently elucidated.

Before we rejoice at this incredible discovery, several questions require our attention:

1) The mass of the oxygen atom is 16 times larger than the hydrogen atom, whereas the letter *yod*, representing oxygen, is much smaller than the *mems*.

2) Modern Hebrew is written or printed with some letters modified when appearing at the end of a word. Final *mem* is one such letter, printed as ם, so that *mayim* is מים, perhaps implying some subtle difference in the two hydrogen atoms in the water molecule. No such difference has been detected to date. Many believe that the special forms of the final letters were only introduced at the time of the Prophets and that the Torah was written with both letters as standard mem. The letter mem has undergone significant modification (as have all the letters of the Hebrew alphabet since ancient times), but the basic bond angle between the letters in the word mayim still maintains the structure of the water molecule regardless.

My granddaughter, Nina Medved, has suggested a mystical connection between the word mayim as written in Hebrew and our discussion of the rakia (Chap. 1c), where God (often written as yod) separated the waters above and the waters below (the two Mems).

With respect to the relative size of the letters and atomic dimensions, the relative size of hydrogen and oxygen atoms is not all that different since the inner shell electrons in oxygen orbit much closer to the nucleus. Fig. IV.5 shows the effect of bond hybridization with a remarkable resemblance to the word mayim. As noted earlier, the repulsion between the residual charges on the hydrogen atoms results in a bond angle of 104.5 degrees, instead of the 90 degrees shown in Fig. IV.5.

Hidden Light: Science Secrets of the Bible

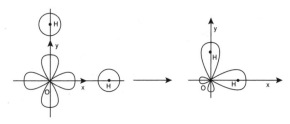

Fig IV.5 – Bond Hybridization – The Structure of the Water Molecule (Ref.IV.13) According to the Theory of "Directed Valence"

3) Concluding remarks

We have seen how the word *mayim* contains at least two secrets of science:

1. It predicts the molecular structure of water with respect to the bond angle.
2. By using the plural form even when applied to the singular entity of water, it anticipates present theories of the multiple molecular clusters responsible for the many anomalous properties of water.

There are also some evocative connections between the molecular weight of water and the *gematria* (numerical value) of Life and the Five Books of the Torah. The atomic weight of oxygen is 16 (8 protons and 8 neutrons) and that of hydrogen is 1 (one proton) so that that the molecular weight of water is 18 – which happens to be the *gematria* of חי (chai – the Hebrew word for life).

Multiplying 18 by 5 gives 90, which happens to be the *gematria* of *mayim* (2 × 40+10).

C. OTHER FORMS OF WATER: ICE AND SNOW

1) The Color of Ice

Liquid water has an absorption peak at 750 nm, as discussed in Section IV.A.5 (The Physics of Color). The absorption peak of ice is downshifted to below 700 nm, as a result of the stronger hydrogen bonding in this solid form of water. Multiple scattering, combined with this shorter wavelength, red-absorption spectrum, results in "…a vivid blueness beneath snow's surface that exceeds in purity that of the bluest sky." – Craig F. Bohren, (Ref. IV.12)

Since tekhelet is supposed to be the color of the "bluest sky (section IV.A.2), one could argue that the author of this beautiful phrase may have been confused. On the other hand, there may be a deeper significance connecting the color of ice and compacted snow to sky-blue tekhelet.

2) Snow in the Bible

Verse 17 of Psalm 147 (which is read every day as part of the morning prayer service) continues the recitation of praises to God's mastery of nature as follows:

"הנותן שלג כצמר כפור כאפר יפזר"

"He gives snow like fleece, scatters frost like ashes"

Fleece (wool) and snow do have some significant similarities. Both are white, serving to reflect direct sunlight. Wool has a remarkable insulation property, resulting from its fiber construction. Loss of heat by conduction and convection is minimized by the structure, which traps the air in the pockets between the fibers. The body heat of someone wearing a wool garment is protected from escaping to the frigid air outside.

An analogous effect is observed with snowflakes, whose many and variegated shapes result in air spaces between the flakes. These air pockets serve to keep the soil under the snow relatively warm in the winter, thus preserving both animal and plant life for the spring awakening.

In Isaiah 1:18, we see another reference to the common properties of wool and snow,

"Come, let us reason together says the Lord – if your sins be like scarlet, they can become white as snow; if they be red like crimson, they can turn white as wool." (also recited during the Musaf prayer on Yom Kippur).

The first five words were frequently used by President Lyndon Johnson when trying to convince his political opponents that he was on the right course.

References – Chapter IV
1. Dr. Baruch Sterman, www.tekhelet.com
2. Dr. Baruch Sterman, "The Meaning of Tekhelet", *B'Or Ha'Torah*, vol. XI: 185–195 [*www.borhatorah.org*]
3. Mendel E. Singer, Ph.D., "Understanding the Criteria for the Chilazon", *Jour. of Halacha and Contemporary Society* 42, (2001) (also available at www.chilazon.com)
4. Darlene Florence, "Spectral Comparison of Commercial and Synthesized Tyrian Purple", *Modern Microscopy Journal* (2003)
5. K.R. Dunbar and R.A. Heintz, "Chemistry of Transition Metal Cyanide Compounds", *Progress in Inorganic Chemistry* 45: 283–391 (1977)
6. Friedlander, et.al., "Uber Brom-Methoxy derivative des Indigos", *Ann. Chem*, 388: 23–49 (1912)
7. J. Formanek, Z. Angew., "Uber den Einfluss Verschiedener Substituenten auf Farbe und Absorption Spectrum des Indigo, Thioindigo und Indirubin", *Chem.* 41: 1133–1141 (1928)
8. Christopher Cooksey, "Tyrian Purple: 66 'Dibromoindigo and Related Compounds", Review Paper in: *Molecules* 6:736–769 (2001) (has 131 references)
9. E. Spanier (Ed.), *The Royal Blue and the Biblical Purple, Argamon and Tekhelet*, Keter Publishing, Jerusalem (1987)
10. J. Wouters and A. Verhecken, "Composition of Murex Dyes", *Journal of Society Dyers Colour* 107 (1991); also 108 (1992)
11. Lionel Milgrom "The Memory of Molecules" [*www.tcm.phy.cam.ac.uk/~bdj10/water.memory/milgrom.html*]
12. Craig F. Bohren, *Clouds in a Glass of Beer*, Wiley, 1987, pp. 155–170
13. Bernard Pullman, "*The Modern Theory of Molecular Structure*", Dover Publications, 1962

Chapter v
Lunar-Solar Periodicities of Large Earthquakes on North-South Rifts

*L*unar-solar periodicities of large earthquakes on N-S faults (San Andreas, Jordan Rift) have been analyzed and observed on both historical and Biblical bases. A statistically significant number of these earthquakes have occurred at new or full moon, sunrise/sunset and within a cluster cell of six years centered on the maximum declination of the 18.6-year lunar orbit. The analysis did not take into account the possible differences in the effects of ocean tides in the Pacific Ocean along the coast of Southern California and in the Mediterranean Sea along the coast of Israel.

1) History of Earthquake Predictions

The early history of earthquake predictions is characterized by scientists studying animal behavior, measuring radon emissions and looking at the night skies for strange lights. As detailed seismic monitoring developed, predictions based on measurement and analysis turned out to be more wrong than correct. At a recent meeting of earthquake specialists, the "p-word" was only whispered or condemned with expletives.

On the other hand, there have been some notable successes. In February 1975, the Chinese government evacuated Haicheng after scientists predicted an impending earthquake based on changes in land elevations, groundwater levels, seismicity and animal behavior. Two days after the evacuation, a 7.3M earthquake struck the area. According to the Chinese authorities, the warning prevented 120,000 injuries and fatalities.

More recently, Prof. Vladimir Kellis-Borok and his colleagues at the UCLA Institute of Geophysics and the International Institute for Earthquake Prediction in Moscow have developed several prediction algorithms using catalogs of historical seismicity (increase in frequency of small quakes, clustering in time and space, etc.) (Ref. v.1). In June 2003, the team predicted a shake of magnitude 6.4 or higher in Central California to occur within nine months. Six months later (in December 2003), an earthquake of 6.5M hit in the southern part of the region.

Another successful prediction, using a method known as RTP (Reverse Tracing of Precursors), is summarized in the following case history:

March 2, 2003: Precursory chain emerged

July 3, 2003: Precursory chain placed on record

September 25, 2003: Tokachi-oki earthquake M= 8.1 occurred as predicted almost six months earlier

Strangely enough, an assembly of 1,224 GPS stations and 1,000 seismometers distributed over the Japanese archipelago failed to give any hint of the Tokachi-oki quake, which had been predicted by the Kellis-Borok team six months earlier. Even more surprising was a quotation attributed to Ichiro-Kawasoki of the Research Center for Earthquake Prediction at Kyoto University: "There was no clear sign at all. It was a shock!"

However, in early 2004, Prof. Kellis-Borok forecast a greater-than-6.3 quake in Southern California, which never happened. The

David Medved

skeptics once again pointed fingers, even though Kellis-Borok had qualified his prediction with a 50% probability that there would be no earthquake. This reservation never made it into public perception.

As a result of public disenchantment, much of the funding has moved from prediction to damage reduction and researchers have begun to speak in terms of forecasting rather than prediction. In the following sections we will employ terms like "the window of maximum vulnerability" which is one step removed from the "p-word."

2) The Kilston-Knopoff Paper on Lunar-Solar Periodicities of Large Earthquakes in Southern California

At 06:01 on the morning of February 9, 1971, ten million inhabitants of the Los Angeles basin (including the author and his family) were awakened by a scale 6.4 earthquake. We rushed out to the deck of our home in the Santa Monica mountains, with a magnificent view of the basin to the southeast, and saw a series of lightning flashes one after the other. My son, Jonathan (age 15) exclaimed: "Dad, it's atomic war!" After some hesitation I responded, "No, sonny, we're seeing the effects of the ground wave tripping the electrical transformers as it rolls south." As we learned later, the epicenter of the quake was in the San Fernando Valley, some twenty miles to the north of our residence.

Another inhabitant of the Los Angeles basin awakened by the same quake, Dr. Steven Kilston (an astronomer with the Hughes Aircraft Company in El Segundo), was quoted by the Los Angeles Times as saying: "I looked out the window and saw a full moon." As the sun was rising in the East, the moon was setting in the West as it always does in the middle of the lunar month (the date of Feb. 9, 1971 corresponded to the 14th day of Shvat on the Hebrew calendar – see Ref. v.2). This observation led to collaboration with Prof. Leon Knopoff, an eminent geophysicist with the UCLA Institute of Geophysics and Planetary Physics.

In 1983, they published the results of their work in *Nature* (Ref. v.3), restricting their analysis to earthquakes greater than scale

Hidden Light: Science Secrets of the Bible

6.0 and to a limited region of the San Andreas fault, between latitudes 33°N to 36°N (San Diego to San Simeon). The San Andreas fault (the meeting place of the Pacific and North American plates) and other major faults in this region run in a roughly NW-SE direction with a significant N-S component. The data is summarized in Table V.1 and the same data is graphically displayed in Figures V.1a and V.1b.

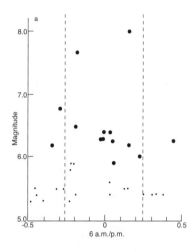

Fig V.1a – Times of occurance of So. California earthquakes on a 12 hour period

Fig. V.1a plots the diurnal variation of Southern California earthquakes (magnitude as a function of time of occurrence) on a scale of 12 hours. Time zero is either 6 A.M. (around sunrise) or 6 P.M. (near sunset), and each division on the time axis represents 1.2 hours. The authors delineated the central part of the diagram to be ± 3 hours from approximate sunrise/sunset times, so that the size of the cell, defined by the vertical dashed lines, is 6 hours (50% of the area in the diagram). The large dots represent earthquakes of a scale greater than 6.0; the small dots are all the smaller earthquakes whose times of occurrence are randomly spread through the 24-hour day. The large earthquakes strikingly "cluster" in the central part of the diagram (10 out of 13 – 77% significantly greater than the area of the cell).

David Medved

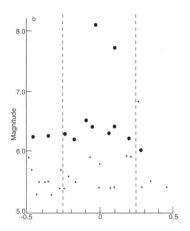

Fig V.1b – Occurance of So. California earthquakes as a function of 18.613 period of lunar orbit. The zero point is 12 January, 1932

Fig. v.1b is based on the 18.613-year periodicity of the precession of the moon's orbit. Put more simply, the moon's orbit tilts northward every year and reaches its maximum declination ("tilt") every 18.613 years. In Fig. v.1b, the zero phase (the northernmost extension of the lunar declination) was chosen as January 12th, 1932 (date of max. declination about one year before a scale 6.3 quake (see Table v.1 below) As with Fig.1a, the large earthquakes cluster in the central portion of the diagram (9 out of 13), whereas the small earthquakes are uniformly distributed. Again, the 69% percentage is larger than the 50% expected on a random basis. The authors also checked the dates of the large earthquakes and found they tended to cluster around the times of a new or full moon, but they did not present the diagram for this case. In Ref. v.3, the authors state "the times of most of the large earthquakes also seem to cluster around the times of new and full moon while the smaller ones are not clustered. The diagram for this case is not presented."

We can get the lunar phase for the dates given in column 1 of Table v.1 by several methods. The Hebrew calendar is a lunar-solar one, based on an average lunation of 29 days, 12 hours, 44 minutes and 3.5 seconds. Lunation is defined as the period of time extending between two successive new moons. The date of the lunar month then tells us which phase of the moon appears on such date with the new moon between the 29th to 30th day of the month and the full moon around the 15th day of the Hebrew month. There are also programs available to determine the moon's phase more accurately on a particular date (Ref. v. 4). All of these need to be used with some caution for dates before 1752, thanks to Pope Gregory and the belated acceptance of the Gregorian calendar by the Western world.. Hebcal, in fact, issues such a disclaimer. As far as can be determined, the dates and phases given in Ref. v.4 are the same as the historical dates listed for the large earthquakes in Southern California, since the earliest data point is in 1857.

Table v.1: Lunar Phase Variations Of Large Earthquakes On The San Andreas Fault

Date	Time (PST)	Mag.	Hebrew Date (Ref. v.3)	Lunar Phase (Ref. v.4)
Mag.>6.0				
9 January 1857	0800	>8.0	13 Tevet – 5617	full
21 April 1918	1432	6.8	9 Iyar – 5678	1Q
22 July 1923	2330	6.25	10 Av – 5683	1Q
29 June 1925	0642	6.25	7 Tamuz – 5685	1Q
10 March 1933	1754	6.3	13 Adar I – 5693	Full, March 11, 1800
25 March 1937	0849	6.0	13 Nisan – 5697	Full, March 26, 1500

David Medved

15 March 1946	0549	6.3	12 Adar II – 5706	Full, March 17, 1100
10 April 1947	0758	6.2	20 Nisan – 5707	3Q
4 December 1948	1543	6.5	2 Kislev – 5709	New, December 1
21 July 1952	0352	7.7	28 Tamuz – 5712	New, 20 July, 1830
19 March 1954	0154	6.2	14 Adar II – 5714	Full, March 19, 0440
8 April 1968	1828	6.4	11 Nisan – 5728	2Q
9 February 1971	0600	6.4	14 Shvat – 5731	Full, February 9, 2341

Using the dates listed in the first column of Table v.1, the Hebrew dates and lunar phases are tabulated in Columns 4 and 5. Column 5 of Table I is derived from the annual tables of Ref. v.4, which list the dates and times (in Universal Mean Time, or UMT) of the new, full and quarter moons. For the San Andreas fault, local time (PST) is 8 hours behind UMT, therefore, the hours given in Column 5 have been corrected accordingly. In this case, 8 out of 13 of these quakes occurred within ± 3 days of either a full or new moon (two at new moon and six at full moon). At 08:00 on the morning of January 8th, 1857, the great quake (M>8.0) hit a sparsely populated area of Southern California. It was shortly after sunrise (middle of winter) and the full moon was setting in the west.

These correlations of large earthquakes with lunar-solar

periodicities on a fault line such as the San Andreas can be summarized as dependant on four conditions:

1. A fault line with a significant N-S component.
2. Either full moon or new moon (± 3 days).
3. Around sunrise or sunset (± 3 hours).
4. In the vicinity of maximum declination of the lunar orbit (± 4.8 years).

These four conditions will be designated as the "KK Conditions." In Sections 3, 4 and 5, these KK Conditions will be tested against major earthquakes on the Jordan Rift and other N-S faults, using the techniques of Kilston and Knopoff (Ref. v.3).

The authors suggest that east-west extensional stress on N-S faults is maximized at moonrise/set and sunrise/set, for lunar and solar tidal influences taken separately, and, consequently, correlations should be maximized at times of full and new moon when the lunar and solar rising/setting times are in proximity. The combined gravitational effects of the sun and moon may release the friction which keeps the plates locked. During a full moon at sunrise (such as in 1971), the sun and moon are aligned on opposite sides of the N-S fault, exerting maximum force on the tectonic plates, pulling them apart. At new moon, the combined pull of the sun and moon is maximized on one plate (at sunrise on the east plate and at sunset on the west plate). The heavenly bodies may not "cause" the earthquake, but their gravitational effects may act as a trigger for an earthquake waiting to happen.

In a recent telephone discussion with Prof. Knopoff (in which I mentioned some of the results from Sections 3 and 4 on the Jordan Rift), he explained that they had not continued this work for several reasons. An attempt to apply a similar analysis to large earthquakes on the northern part of the San Andreas failed to give similar correlations. Both in their article and subsequent newspaper stories, these scientists clearly stated that they were not making any predictions. However, based on the clustering of large earthquakes over a period of 18.613 years (Fig. 1b), their paper stated "...during a window of a few years which astride this date (November 1987), one might be more likely than otherwise to observe one or more large earthquakes in Southern California." However, not much happened until 1994, far from their "window of maximum vulnerability." As a result of these and other factors, their paper was largely dismissed and forgotten.

3) Earthquakes in the Bible (The Jordan Rift)

There are many references to large earthquakes in the Bible (Ref. v.5). The land of Israel is on the Africa plate with the Jordan Rift as the interface between it and the Asia plate. Running almost due N-S throughout its length (the Hulah, Sea of Galilee, Jordan River Valley, Dead Sea, Arava and into the Gulf of Eilat), this fault becomes the Great Rift Valley in Africa. It is quite remarkable that a number of these Biblical accounts fit the lunar-solar periodicities of the KK conditions, 3,600 years prior to the publication of their paper.

Consider the most striking example: The destruction of Sodom and the cities of the Jordan Plain in the year 2047 (1713 BCE). The area is described as a subtropical paradise in Genesis XIII:10, but populated by very disreputable people. God tells Abraham (Genesis XVIII-20): "Because the cry of Sodom and Gomorrah is great and because their sin is very heavy – extermination..." Two angels are sent to carry out the mission. They first visit the 99-year-old Abraham in Hebron to tell him that the following year, his wife Sarah, at 90 years of age, will bear him a son. By citing their ages, the Bible informs us that these events occurred in the year 2047. The angels then proceed to Sodom and are hosted by Lot (Abraham's nephew) who bakes them unleavened bread, or *matzot*.

We are told by Rashi (and others) that these visits were on the eve of Passover, the 14th-15th of Nissan.[7] After much urging, Lot escapes from Sodom with his wife and two unmarried daughters, at just about the time that dawn's first light was streaking the eastern

---

[7]. The basis for this conclusion, and an explanation of the Patriarchs' observation of the customs of Passover, 400 years before the occurrence which the festival commemorates, are not stated.

sky (Genesis XIX:15). The next lines (Genesis XIX:23–25) struck this writer like a thunderbolt, since a few months earlier I had read the Kilston-Knopoff paper: "The sun rose upon the earth when Lot arrived at Zoar. God (then) rained upon Sodom and Gomorrah brimstone and fire from the hand of God from the heavens. And he overturned these cities and the entire plain and all of the inhabitants of these cities..." The Hebrew word for overturned – ויהפך (VaYahafoch) literally means "turned them upside down," a most graphic description of a great earthquake, that, together with volcanic eruptions, transformed in a matter of moments a verdant valley into the Dead Sea and its barren surroundings of today.

All four of the KK conditions were operative:

1. At Passover the moon is full.
2. The earthquake occurred at sunrise with the full moon setting in the west.
3. The Jordan Rift at the Dead Sea runs in a straight N-S direction.
4. The lunar orbit was three years short of the maximum 18.613-year declination, well within the cluster cell defined in Fig. v.1b.

The next example (Joshua and the walls of Jericho) was suggested as a possible biblical fit by Professor Knopoff.

According to the Bible, the Exodus of the Israelites from Egypt occurred on the 15th day of the lunar month, Nissan, in the year 2448 in the Hebrew calendar (1312 BCE). Following the forty years of wandering in the Sinai desert and the death of Moses, Joshua led them across the Jordan River into Israel on the 10th of Nissan, 2488 (1272 BCE). This second crossing was quite miraculous since the river at this time of year is usually at flood tide (at least 30 meters across and up to 4 meters deep), being fed by great springs in the far north and the melting snows of Mt. Hermon. Some authorities cite landslide blockages of the Jordan upstream as caused by earthquakes, similar to those that occurred in more recent times (1160, 1534, 1906, etc.).

Such an earthquake at the time of Joshua's crossing may have been a precursor to the great shake which followed a few weeks later (see footnote 16 on page 16 of Ref. v.6): The Children of Israel encamped at Gilgal in the plains of Jericho following the crossing and their mass circumcision at Givat HaOraloth (the hill of the foreskins). They kept the Passover on the evening of the 14th day of Nisan (Joshua v: 9–10).

Many archaeologists and Biblical scholars are in agreement that the walls of Jericho were not breached, but collapsed as from the effect of an earthquake – the Hebrew language in Joshua VI:5 is strikingly evocative: "...יריעו כל העם תרועה גדולה ונפלה חומת העיר תחתיה" (...the people shall shout with a great shout and the wall of the city shall collapse upon itself). These were God's words to Joshua and this is exactly what occurred on the seventh day after seven circuits of the city walls by a strange procession of seven priests blowing seven horns, followed by the Ark of the Covenant and a completely silent army in front of it all.

"ויהי כשמע העם את קול השופר ויריעו העם תרועה גדולה ותפל החומה תחתיה" –
"And it came to pass, when the people heard the sound of the horn that the people (nation) shouted a great shout and the wall collapsed upon itself" (Joshua VI:20).

There is no direct indication in the entire Book of Joshua as to the date or the time of day when the wall collapsed. However, there are several clues in the language which point to an answer. At least 14 days elapsed between the celebration of Passover Eve at Gilgal and the fall of Jericho (7 days of Passover plus 7 days of marching around the walls – 1 circuit each of the first six days and 7 circuits on the seventh day). This brings us to the 29th of Nissan – almost new moon. In fact, there is an 18th century text, *Seder Hadorot*, written by Rabbi Yechiel, where he states that the destruction of Jericho occurred on *Rosh Hodesh* Iyar (the first day of the month of Iyar) (Ref. v.7). According to this statement, the Children of Israel took two days of preparation after Passover before commencing their fateful circumnavigations of the city.

Hidden Light: Science Secrets of the Bible

There is a tradition that the chief god of Jericho, or, in Hebrew, יריחו, was the moon god. This should not be too surprising, since the Hebrew word for moon is ירח (yareach), as in יריחו. Choosing the day of the invisible new moon (*Rosh Hodesh* Iyar) as the day of its destruction would be poetic justice. However, Rashi points out that Sodom was destroyed when both moon and sun were in the sky (see his commentary on Genesis XIX:25), thus equally punishing the idolatrous worshippers of these heavenly bodies.

A few years ago, the Palestinian Arabs held a festival in honor of their descent from the ancient Canaanites. Although many scoffed at these assertions, made by second and third generations of descendants from those who came to Israel from other areas in the 1930's (mostly Syria and Egypt), serious scholars should study the possible connection, at least on a spiritual level, with the hapless residents of Jericho.

The pagan Arabs of Mecca and Medina worshipped Allah the moon god as their chief deity, and, after becoming Moslems, still maintained a strong lunar orientation. Their calendar is strictly lunar, with no solar correction, and their banner was the crescent moon, which even today appears on the flags of all 22 Arab countries.

Even more intriguing is the determination of the time of day when the earthquake struck and caused the walls to collapse. After the description of the first circuit on Day 1, we read in Joshua VI:12: "וישכם יהושע בבוקר" (And *Joshua* arose early in the *morning*...), and the same process for Days 2–6 is repeated. However, on Day 7 (VI:15): "...ויהי ביום השביעי, וישכמו בעלות השחר" (And it came to pass on the seventh day that *they* rose early at the *break of dawn*) and immediately started the seven circuits. There are two questions which cry out for an answer:

1. How much time did it take to encompass the city on one circuit (or how long was the perimeter)?
2. How much time elapses between crack of dawn and sunrise at that time of year?

At a recent family gathering, I asked my niece's husband, a well-known Israeli archaeologist, the first question. Without any hesitation he answered: "Ten minutes." When I expressed astonishment and disbelief, he explained that the "cities" of those days were at most 200 meters on a side. Thus, the seven circuits would take about 70 minutes.

There is a disputation on the answer to question 2 amongst the rabbis. They all agree that the time between the first dawn and sunrise is the time it takes a healthy man to walk 4 mil (a Talmudic unit of about 1 km.). The disputation revolves around the speed of his walk, varying from 18 minutes per mil to 22.5 minutes. The spread is therefore between 72 to 96 minutes. As an experimental scientist living in Jerusalem (the same latitude as Jericho), I decided to check it personally and measured about 72 minutes (in early May). Putting these answers together would imply that the earthquake which caused the wall to "collapse upon itself" occurred at sunrise.

The analogy with the lapse of time between dawn and sunrise in the great earthquake which destroyed Sodom is quite striking. Lot left Sodom at first light and arrived at his refuge (Zoar) at sunrise (Genesis xix:15–23).

Once again the four KK conditions were simultaneously operative in Jericho:

1. N-S fault (practically ground zero)
2. New moon
3. Sunrise
4. Two years past maximum declination (well within the cluster cell)

There was a major earthquake during the reign of King Uzziah in Jerusalem (52 years, or 786–734 BCE). However, it was not possible to trace the date, time of day, or even the year in the sources. The reign of King Uzziah was marked by a revival of both military and religious fervor in Judea, but it was flawed by an inexplicable change in the King's character after a string of military and other successes.

In Chronicles 2 XXII:1–12 there is no reference to the earthquake which struck during the reign of Uzziah. However, in the Book of Prophets there are two references:

1. The first lines in Amos 1:1 state: "The words of Amos, who was among the herdsmen of Tekoa...in the days of Uzziah, King of Judah...two years before the earthquake."
2. Zechariah XIV:5: "...ye shall flee, like as ye fled from before the earthquake in the days of Uzziah, King of Judah.."

As some commentators have noted, earthquakes were not uncommon on the Jordan Rift, so the one mentioned in these two sources must have been of unusual severity. The only detailed account of these events is provided by Josephus in his *Antiquities of the Jews* (Ref. v. 8):

> "...he [Uzziah] was corrupted in his mind by pride...and accordingly, when a remarkable day was come and a general festival was to be celebrated, he put on the holy garment and went into the Temple to offer incense to God upon the Golden Altar..."

Ancient Israel, even before the destruction of the First Temple, firmly adhered to a doctrine of separation of Church and State. The Kings of Judea, like Uzziah, were descendants of the House of David, whereas the *Kohanim* (Priests) were descendants of Aaron. Consequently, Uzziah was confronted by a horrified Azariah (the *Kohen HaGadol*, or High Priest) and a cohort of 80 others who pointed out that "none besides the posterity of Aaron was permitted to do so."

The king became furious and threatened to kill them all. A great earthquake then shook the ground and a tear was made in the Temple. The bright rays of the sun hit the king's face and he was stricken with leprosy. He was led out of the city and lived the remaining 17 years of his life as a recluse while his son Jotham took over the government and did well. Upon the death of Uzziah in 734 BCE, Jotham became king in name as well as in practice.

Accordingly, the year of the great earthquake can be determined as 751 BCE (2.7 years after maximum declination).

We can only speculate on the time of day and lunar phase – as usual, such speculation is guided to some extent by the answers one is seeking. There are two clues in the Josephus account: First, "…a remarkable day was come and a general festival was to be celebrated": It may have been any one of the three Pilgrim festivals: *Pesach* (Passover), *Shavuot* (Pentecost) or *Succot* (Tabernacles). The first days of *Pesach* and *Succot* are always full moon.

The second clue, "…he went into the Temple to offer incense on the Golden Altar," refers to the daily incense offering, usually served up in the early morning and in the evening. The evidence that the KK Conditions were in force at this tragic event is suggestive only and will not be included in our tabulation of large earthquakes on the Jordan Rift.

In the following section we will review the historical records of large earthquakes on the Jordan Rift from 31 BCE to the present day.

4) Large Earthquakes on the Jordan Rift since 31 BCE

In Table 11 we have tabulated nine earthquakes recorded on the Jordan Rift in the period from 31 BCE to the current era (1995). The data for the time of day was not available for most of the early ones and the date of September 2 for the great earthquake of 31 BCE is based on Josephus' *Antiquities* (Ref. v.9). Josephus writes that in the seventh year of the reign of King Herod (and around the same time that the Battle of Actium raged between the forces of Marc Antony and Octavius), there was an earthquake in Judea – "such a one as not happened at any other time and which earthquake brought a great destruction on the cattle in that country. About 10,000 men also perished by the fall of houses." What an insight into the mind of Josephus – he first laments the fate of those poor cattle. This is much worse than the plea of Reuven and Gad in the Book of Numbers, which evokes a sharp reprimand from Moses. They had asked to build shelters for their livestock first and then houses for their families, and Moses promptly reversed the order.

The lunar phase data shows that five out of nine of these

quakes occurred either at full moon or new moon, with some skew toward the new moon. The orbit declination data show some clustering within a cell less than plus/minus two years wide (four out of nine) and five out of nine in a cell the size of the Kilston-Knopoff paper. In only two cases are all of the KK Conditions simultaneously fulfilled on a given earthquake date (1202 and 1856).

The data on diurnal occurrences of the earthquakes listed in Table II has not been readily available in most cases. A notable exception is the earthquake of 1202 (Ref. v.10). A remarkable collaboration of experts in seismology, archaeology and history studied and analyzed the ruins of a Crusader castle, called Vadum Iacob, 3 km. south of the Huleh Valley on the Jordan Gorge. The castle was besieged and conquered by Saladin in 1179, just 23 years before an earthquake of M>7 hit at dawn on May 20, 1202. The castle walls were offset by 1.6 meters.

The probability, P, of simultaneous occurrence of the three conditions (lunar phase, orbit declination and proximity to sunrise/sunset) on a random basis depends on the sizes of the cells. For lunar phase choose ± 3 days, orbit declination ± 3.5 years and proximity to sunrise/sunset ± 3 hours, then $(Pd=6/29.5)(Pl=7/18.613)(Ph=6/24)=P$, where Pd is the random probability of an event occurring within a cell 6 days wide out of the 29.5 days of lunation, Pl that the event will occur within a cell 7 years wide centered around the 18.613 year maximum of lunar orbit declination, and Ph that the event happens at a time of day within 3 hours of sunrise or sunset. The 4th KK condition is, of course, a fault with a significant N-S component, which does not constitute a variable to be included in these calculations. On a random basis, the probability of a simultaneous convergence of all four KK conditions is $P=0.019$, or roughly 2%.

Two out of nine events per Table II (below) would represent a 22% actual occurrence. If we add the two Biblical events (Sodom and Jericho), where all KK conditions were operative simultaneously, the result is 36% (four out of eleven) of the major earthquakes on the Jordan Rift occurring at the simultaneous convergence of the KK conditions, compared to a random non-correlated group which would give 2%.

Table 11: Major earthquakes on the Jordan rift – historical record

Date		Lunar Phase	Time to/from Max. Declination		
Year	Month & Day	Ref. v.4 & 5		Time	Magnitude
31 BCE	Sept. 2	2 days after new moon	9.3 years	– –	> 7.5
746	Jan. 18	Last quarter	0.4 years	– –	7.3
1157	Aug. 15	First quarter	0.0 years	– –	> 7.0
1202	May 20	3 days before new moon	3.5 years	~ 0530 Dawn (see Ref.v.10)	7.5
1759	Nov. 25	First quarter	5.3 years	– –	7.4
1837	Jan. 1	Last quarter	1.9 years	– –	6.5
1856	Dec. 12	Full moon	1.5 years	TBC	TBD
1927	July 11	3 days before full moon	5 years	– –	6.8
1995	Nov. 23	1 day after new moon	7 years	0415	7.5

5) The Great Indonesian Earthquake and Tsunami of December, 2004

On December 26, 2004 at 06:58 (local time), a massive earthquake (magnitude greater than 9.0) erupted beneath the ocean floor near the west coast of Sumatra. Apparently, the epicenter was on the fault line formed at the interface of the Indian-Asian (Burmese) tectonic plates.

On the Indian subcontinent the fault runs essentially in an east-west direction (along the Himalaya mountain range). As it enters the Bay of Bengal, it veers sharply to the south and at 5 degrees N latitude, 95 degrees E longitude (corresponding to the estimated epicenter), the fault has a N-S orientation.

On the morning of December 26, 2004, there was a full moon and the quake occurred as the sun was rising and the moon was setting. We were less than 1.5 years from the maximum northward orbital declination of the moon, expected to occur between April and June of 2006 (between *Pesach* and *Shavuot* of 5766). All four KK conditions were thus in convergence.

6) Conclusions and Forecasts

We have noted statistically significant correlations of lunar-solar phases with large earthquakes on two faults with a significant N-S component, and cited a similar correlation with the recent Indonesian earthquake. As noted above, we are still well within "the window of maximum vulnerability" which extends for 4.8 years past the maximum of April-May, 2006 (to November-December, 2010).

It is suggested that there should be increased and intensive monitoring (using seismic and optical methods) of the Jordan Rift in a cooperative endeavor between Israel and Jordan, with significant international support.

David Medved

List of Figures

Fig. v.1a Times of Large Earthquakes in Southern California (day/night variation)

Fig. v.1b Times of Occurrence on a Scale of Period 18.613 Years (lunar orbit)

Table v.1 Major Earthquakes in Southern California by dates and lunar phase

Table v.2 Major Earthquakes on the Jordan Rift since 31 BCE

References – Chapter v

v.1 P. Shebolin, v. Kellis-Borok, et. al., "Advance Short-Term Prediction of the Large Tokachi-oki Earthquake, September 25, 2003, $M = 8.1$. A Case History," *Earth Planets Space* 56:715–724 (2004)

v.2 *www.Hebcal.com/converter*

v.3 S. Kilston and L. Knopoff, "Lunar-Solar Periodicities of Large Earthquakes in Southern California," *Nature* 303: 21–25 (1983)

v.4 Quick Phase Pro [http://www.calculatorcat.com/moon-phases/]

v.5 Dr. Lambert Dolphin in *www.ldolphin.org.quakes.html*, original article written Feb. 1994, updated Mar. 2004

v.6 Dr Rudolph Cohen (ed.), *Joshua and Judges*: Soncino Books of the Bible: footnote on verse 16, page 16

v.7 Rabbi Yechiel of Minsk, *Seder Hadorot*, Chapter 18 (based on the 4th and 5th sections of *Seder Olam Rabah*)

v.8 Josephus Flavius. *Antiquities of the Jews*, Book IX, Chap. x:4

v.9 *IBID*, Book XV, Chap. v

v.10 Vadum Iacob Research Project [http://vadumiacob.huji.ac.il/]

Chapter VI

Archaeology and the Bible (the Exodus from Egypt)

There is continuing controversy in the archaeological community concerning the date of the Exodus. Various arguments based on scant evidence present a spread of 300 years for the departure of the Children of Israel from their Egyptian bondage (from 1500 to 1200 BCE).

In this chapter, a tentative date is established by employing a combination of archaeological evidence, biblical narrative and commentary with astronomical observations.

1) Open Questions on Exodus

Biblical scholars, historians and archaeologists have long been enmeshed in a running debate on the events of the Exodus. The date, the route and the Pharaoh at the time of the Exodus are still unresolved. The early Christian Church was racked by sharp dissension on these questions. Some authorities even claim that the whole story is a fiction, citing the lack of any evidence on two accounts:

1. No traces have been found in the sands of the Sinai Peninsula to show that two million Israelites traversed and camped in this wilderness (with the possible exception of Kadesh Barnea (Ref. VI.1).

2. None of the Egyptian monuments and hieroglyphic records mentions the plagues, the departure of the Israelites or the disaster at the Red Sea. The Merneptah Stele (Ref. VI.2), which contains the earliest known reference to Israel outside of the Bible, identifies it as a people already resident in Canaan.

The dates assigned to the Exodus vary from the fifteenth century BCE to the early thirteenth century BCE (1500 to 1200 BCE). The route of the Exodus is also a matter of conjecture. A famous archaeologist recently confided to this author that anyone who attempts to rationalize and resolve the debates on archaeological dates is entering a swamp. James Michener, an American author, wrote in 1979: "Conventional chronologies have served us long enough and not too well as an interim tool. Most tools need sharpening over the years and final replacement." He was eminently qualified to make such a critique, composing sweeping sagas spanning many generations at a given location following meticulous historical research, such as "The Source", set in Israel from prehistoric to modern times.

So, with much humility and buoyed by the life preserver of the Bible, let us prepare to enter the swamp.

2) The Date According to the Bible

The chronology of events described in the Five Books of Moses is well-established. For example, the birth of Abraham occurred in the year 1948 of the Hebrew calendar (1812 BCE) and Isaac was born in 2048 (1712 BCE), when Abraham was 100 years old (Genesis XXI:5).

On the other hand, the Bible has contributed to the confusion about the date of the Exodus. In Parashat *Boh* (Exodus XII:40–41), we read that "...*the children of Israel resided in Egypt for four hundred and thirty years. And it came to pass that at the end of four hundred and thirty years, all the Hosts* [Legions] *of God left Egypt in broad daylight.*" This number is at variance with the prophecy given by God to Abram at the Covenant of the Parts (Genesis XV:13): "...*know for sure that your descendants will be strangers in a land not their own and shall serve them, and they shall afflict them for four hundred years.*" Even more serious is the calculation by Rashi and other commentators on Israel's actual residence in Egypt, which starts with the descent of Jacob and his family to Goshen. Only five generations separated Jacob from Moses. The Bible states that Jacob was 130 years old when he appeared before Pharaoh and that Isaac was 60 years old when Jacob was born. There is general agreement that the four hundred years predicted in Genesis XV date from the birth of Isaac. If we take the birth of Isaac in 2048 as the defining beginning of the wandering and exile, then the Exodus occurred 400 years later, in 2448 (1312 BCE).

Therefore, the Israelites sojourned in Egypt for 210 years (400–130–60).

The 430 years of Exodus XII pose much bigger problems. Some of the commentaries claim that the Covenant of the Parts occurred when Abram was 70 years old (before he departed Haran). Since Isaac was born when Abraham was 100 years old, this adds thirty years to the 400 years.

Question: If the 430 years date from the Covenant of the Parts, then why did God tell Abram that the exile would last 400 years.

Possible Answer: Since *Hashem* also promised Abram in the same passage that he would have descendants, the exile was dated from the birth of Isaac, his son.

The Septuagint tries to dodge the issue by intentionally mistranslating the sentence (Exodus: XII:40), in stating: "...the Children of Israel resided in Egypt *and Canaan* for four hundred and thirty years."

There are two other sources in the Bible which have been

cited for determining the date of the Exodus. In the Book of Judges we are told of the war between the Israelites and Ammon. The King of Ammon laid claim to the territories (in TransJordan) captured by Israel from the Amorites during the last years of the Exodus (in 2488 – see Numbers XXI:21–35). Jepthah was selected by the Elders of Gilead to be the Judge and the General of the Israelite forces. Before engaging the Ammonite army in battle, Jepthah sends a long conciliatory message to the King of Ammon, giving him a detailed history lesson. The key sentence in Judges XI:26 states: *"While Israel dwelt in Heshbon and its towns.. and in all the cities that are along by the side of the Arnon three hundred years; wherefore did ye not recover them within that time?"* According to Rashi, this message was sent around 2788, or approximately 300 years after Joshua. Since 40 years elapsed from the Exodus to the death of Moses, the date of the departure from Egypt, is once again, 2448. Parenthetically, the King of Ammon rejected the olive branch and prepared a large army to attack Israel. Under the inspired leadership of Jepthah, the Israelites dealt the Ammonites a crushing blow, *"…so the children of Ammon were subdued before the Children of Israel"* (Judges XI: 33).

Early in the First Book of Kings, we learn of Solomon's decision to begin construction of the Temple in the fourth year of his reign: *"And it came to pass that four hundred and eighty years after the Exodus of the Children of Israel from the land of Egypt, he* [Solomon] *began the building of the House of God"* (1 Kings, VI:1). According to the chronological tables of Eliezer Shulman (Ref. VI.3), the year was 2928 BCE, leading once again to the year 2448 as the time of the Exodus.

There is a serious discrepancy of 134 years between the date of 2928 and the year 966 BCE, assigned by most secular authorities to the commencement of construction (see Section VI.E on dating problems).

3) The Route According to the Bible
The opening lines of Parashat *B'Shalach* (Exodus XIII:17) describe the route which was *not* taken by the Children of Israel:

"ויהי בשלח פרעה את העם ולא נחם אלהים דרך ארץ פלשתים כי קרוב
הוא כי אמר אלהים פן ינחם העם בראתם מלחמה ושבו מצרימה"

> "And it was, when Pharaoh sent out the people, God did not lead them by way of the land of the Philistines, although it was close; for God said lest the people lose heart when they see war and return to Egypt"

The astute reader could reasonably ask a number of questions on this long sentence:

(1) *"Because it was close"* – Why in the world would a benevolent and loving God take the Children of Israel on a circuitous route (southeasterly), when the direct way along the coast was so much shorter?

(2) What is this "war" that they might encounter?

(3) Who are these "Philistines?"

There are many commentaries and *Midrashim* (none of which is particularly satisfactory) that try to explain the hidden meanings in these words.

1. The route: Rashi, for example, states that the very closeness of the coastal route would make it all too easy to return to Egypt at the first sign of trouble.

2. The threat: Secular scholars explain that Egypt, as the great imperial power of the period, had constructed a chain of forts all along the coast. Bas reliefs at Karnak (on the north side of Luxor) show such a string of fortifications along the Way of Horus (the Egyptian name for the coastal route). High bastioned citadels and adjoining reservoirs are also inscribed in detail on the Temple of Aman at Thebes, apparently in commemoration of the victorious campaign

of Seti I (father of Ramses II) against Canaan (see an excellent and informative article recently published in BAR – Biblical Archaeology Review – Ref. VI.4).

Another explanation comes from a *Midrash* which tells how *Bnai Ephraim* could not wait for God's promised deliverance and escaped from Egypt 30 years before the Exodus, traveling along the Philistine Highway. According to the Talmud Sanhedrin, they incorrectly reckoned the 400 years as dating from the Covenant between the Parts. They were attacked and mercilessly slaughtered. Their bleached bones were strung along the route, a sight which certainly would have triggered a quick return to Egypt.

As it turned out, the route traveled by the Israelites did not help them avoid war. In fact, they engaged in two different "battles" almost immediately after leaving Egypt. Just seven days into the Exodus, Pharaoh and his army caught up with them and trapped them on the shores of the Sea of Reeds (not the Red Sea, as mistakenly translated from the Hebrew *Yam Soof*). There may have been skirmishes between the two camps during the night, until the sea was "driven back" by a powerful east wind.

Following the drowning of the Egyptian forces, the people traveled to Rephidim (near Mt. Sinai), where they were attacked by Amalek. The attack was entirely unprovoked, since they were a great distance from the territory of Amalek (near Petra in TransJordan). In an all-day battle, the Israelite forces under Joshua were victorious. Moses was instructed by God to write these words in the Torah: "...*God shall be at war with Amalek for all generations.*" So, is this the way they avoided "war," by not taking the coastal route? Several explanations have been offered:

a. They were destined to witness The Revelation at Mt. Sinai.
b. In both cases they were attacked in relatively remote areas, where a quick and easy return to Egypt was not possible.

3. Who were the Philistines (literally meaning "Invaders")? There is general agreement that the "Pelishtim" of Genesis XXI and XXVI are not the same Philistines who confronted Samson, Saul and

David. There are many Biblical references to the Philistines, one particularly relevant to our discussion. In the historical review given by Moses in his hortatory address to all of Israel (Deuteronomy 11: 17–23), he describes how the Ammonites (descended from Lot) and the Edomites (descended from Esau) annihilated the original inhabitants and dwelt in their lands. In sentence 23 he states: "*And the Avvim who dwelt in Hazerim, as far as Gaza; Kaftorim who came from Kaftor, destroyed them and dwelt in their stead.*" The Bible tells us that 40 years after the Exodus (2488) invaders from Crete (Kaftor) were already well established in the "Land of the Philistines." Note that this disclosure is dated about 400 years after the sojourning of Abraham and Isaac at Gerar (between Gaza City and Rafah). In the following section, we will attempt to provide linkage between 200 years of archaeological discovery and these Biblical verses.

4) Archaeological Evidence

In 1798, Napoleon Bonaparte led a large French army on an invasion of Egypt, commencing with an amphibious assault on the beaches near Alexandria. In the spirit of the French revolution, he took along 167 scholars and scientists to explore the antiquities.

Fig V.1– Medinat Habu bas relief showing naval battle between Egyptians and Sea Peoples

Following the defeat of Mehmoud Bey and his Mameluke troops at the Battle of the Pyramids, the remnant escaped toward the reaches of the Upper Nile. A French expeditionary force chased after them. Dominique Vivant Denon, an artist and man of letters, accompanied the troops as far as the temple complexes of Luxor and Thebes, where he made some remarkable discoveries. At the Temple of Medinet Habu, near Thebes, he found a series of wall carvings showing the Egyptians engaged in ferocious land battles and naval engagements with strange enemies (Fig. VI.1). Denon meticulously recorded these scenes, but was unable to decipher the hieroglyphics. Some decades later, Jean Francois Champollion and others recognized that the battle inscriptions gave an account of a furious defense of Egypt against invaders from the sea, during the reign of Ramses III (1190 to 1165 BCE). By the end of the nineteenth century it was established that the captives wearing the feathered headdresses (Fig. VI.1) were the 'Sea Peoples', including the Philistines. In "People of the Sea – the Search for the Philistines," it is stated that "...the invasion of Egypt by the Philistines and their allies was only the last phase in a wave of destruction that had swept through most of the known world" (Ref. VI.5). The landing of these Sea Peoples on the shores of the eastern Mediterranean and their subjugation of the indigenous inhabitants from Phoenicia to Gaza could be considered the first "Hundred Years War," extending through the fourteenth and thirteenth centuries BCE. The way of the Philistines, passing through the Nile Delta and along the coast, would have been the scene of battles like those depicted in the Medinet Habu wall carvings. Is this the "war" that was to be avoided by the Children of Israel? According to the recitation of Moses in Deuteronomy 11:23, a branch of the Sea Peoples had already replaced the original inhabitants on the southwest coast of Canaan at the time of the Exodus.

These are the Philistines who became the mortal enemies of the Israelites. It is not surprising that the two peoples had a violent collision. The Philistines tried to conquer Canaan, invading it from the west, whereas the Children of Israel came in from the east. The Canaanites didn't have a prayer.

David Medved

5) Dating Problems

At the conclusion of Section 2, it was noted that there exists a significant difference between the dates assigned to the Exodus by Biblical scholars and those accepted by most secular historians.

The Biblical year 2928, given by E. Shulman (see Ref. VI.3) as the year Solomon initiated the building of the Temple, corresponds to 832 BCE, whereas the Encyclopedia Judaica (Vol. 15, pp. 96–98) lists it as 966 BCE. In order to arrive at the latter date, scholars have started with the year 586 BCE, accepted as the time of the destruction of the First Temple (based on Babylonian records). They then worked backwards, adding up the years of the Kings of Judah, including Solomon (385 years). Working back another 480 years gives 1446 BCE as the date of the Exodus. This result is achieved by combining the Biblical narrative in the Book of Kings with archaeological and historical records (Assyrian, Babylonian). Unfortunately, the evidence is not conclusive and this date is still subject to controversy.

We have, throughout our study, used one of the most effective tools of Biblical exegesis; namely, to ask where else does the Bible provide us with clues to achieve a solution of the problem? Consider the following question: Are the years listed in the Bible lunar or solar years? There are several portions in the Bible which may have addressed this question. The moon completes its orbit around the earth in 29 days, 12 hours and 44 minutes (the mean synodic lunar month). Therefore, twelve lunar months (a lunar year) are about 11 days less than the solar year of 365.25 days.

Over the last two or three millennia, the Hebrew calendar has used a combination of lunar and solar cycles, together with intercalation (insertion of extra days), similar to the correction in the Chinese calendar. There is a nineteen-year cycle where 7 out of the 19 years have an extra month of 30 days inserted (Adar II), to make up the approximately 210-day shortfall of the lunar years. It is not known from historical records when the people of Israel may have accepted this calendar.

On the other hand, the Torah is quite clear on this issue. In *Parshat Emor* (Leviticus XXIII:4), *Hashem* tells Moshe:

"אלה מועדי ה' מקראי קודש אשר תקראו אתם במועדם."

"These are the festivals of the Lord, holy gatherings, which you shall proclaim in their appointed seasons." In the Book of Deuteronomy (XVI: 1) we are instructed:

שמור את חדש האביב ועשית פסח לה, אלהים כי בחדש האביב הוציא ה' אלהים ממצרים לילה

"Observe that the holiday of Passover occurs during the month of the springtime since it was in the spring month that *Hashem*, your God brought you forth from the land of Egypt at night"

In other words, we were commanded at the giving of the Torah at Mt. Sinai to carry out the process of intercalation. When was that? Exactly 50 days after the Exodus in the year 2448. What about the times before the Revelation at Sinai? How are those years counted?

In the year 2047, when Abraham was 99 years old, he is visited by three "wayfarers" who have come to announce that he will have a son. Instead of saying "next year" (בשנה הבאה) they tell him בעת חיה, literally "at the living season." Rashi and other commentators state that one of the "visitors" put a mark on the wall, indicating that when the sun hits the mark the following year, Abraham's son will be born.

At this point, let us speculate. If the pre-Sinai years, as given in the Bible, were indeed only lunar years, then in the 2,448 years from Creation, only 2,374 solar years would have elapsed. Thus, the lunar year 2448 is 74 years ahead of the solar year 2374. In such a case, the Exodus would have occurred in 1386 BCE. This leaves a discrepancy of 60 years with the secular date of 1446 BCE.

6) Is Evil (רעה) MARS?

As noted in the previous section, archaeologists have employed ancient records of astronomical events to support their findings and conclusions. In recent years, a new interdisciplinary approach called archaeoastronomy has developed, originally stimulated by the question: Could ancient man have built stone structures (like those at Stonehenge) to measure…movements of sun, moon and stars? Our approach here could be called Biblical Archaeoastronomy, as it combines biblical exegesis with astronomical observations and analysis in seeking solutions to archaeological questions. Consider the following example, where the Bible provides an astronomical clue to the date of the Exodus.

In *Parshat Boh* (Exodus x:10), Pharaoh, having just recovered from the seventh plague (hail), tells Moses and Aaron:

"ויאמר אלהם יהי כן ה' עמכם כאשר אשלח אתכם ואת טפכם ראו כי רעה נגד פניכם."

"And he [Pharoh] said to them, Let the Lord be so with you, as I will let you go, and your little ones; look out for evil is before you." Rashi quotes a Midrash on this strange sentence: "There is a certain star, the name of which is evil (Ra'ah). Pharoh had said to them: "By my astrological art, I see that star rising toward you in the wilderness whither you wish to proceed. It is an emblem of blood and slaughter." Other commentators say that it may have represented the Egyptian sun-god Ra'ah.

Following Rashi, this description is certainly evocative of the red planet Mars, rising in the evening skies. Apparently Pharoh thought that the Children of Israel would be marching eastward as they left Egypt with this Ra'ah shining directly in their faces

A recent article by John Rummel (Ref. VI.6) employs remarkably similar language to the Midrash quoted by Rashi. Published a few months before the closest approach of Mars to Earth in almost 60,000 years, he writes:

"Mars is most often an inconspicuous orange starlike object that wanders eastward among the stars. But about every two years, it begins to brighten, reverses its course and is prominent as a fiery orange red star in the evening sky. At times, it even exceeds mighty

Jupiter in brightness. Then after a month or two it begins to fade and resumes its anonymity among the starry host."

Clearly, if we can identify those times when Mars would be transformed into such a fiery red star, we might have a better handle on the date of the Exodus.

Consider Fig.VI.2: Planets closer to the sun than their neighbors have faster orbital velocities (Kepler's Law). Therefore, the Earth will overtake Mars about every 26 months. These close encounters, or Mars oppositions, are particularly noteworthy when they occur close to the Martian perihelion (that point in its elliptic orbit when it's closest to the sun). These particular encounters are designated 'perihelic oppositions'.

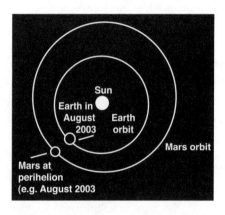

Fig VI.2 - Perihelic Oppositions
(Close Encounters Between Earth & Mars)

The eccentricity (deviation of the elliptic orbit from circular) of the Martian orbit is much greater than that of Earth (about 5x). Each opposition is followed by a very similar one 79 years later (Ref. VI.7). For example, the opposition of Aug. 28, 2003 was a repetition of the oppositions of August 23, 1924 and of August 18, 1845. Over the years, the distance between the planets will vary from 56 million to 400 million kilometers, resulting in large changes in the size and brightness of the Red Planet as it appears to Earth.

In response to my inquiry, Dr. Stephen McCluskey of West Virginia University and the Institute of Archaeoastronomy has calculated the dates of Mars-Earth oppositions in Universal Time (UT) and the separations of the planets in Astronomical Units (A.U.), using the Skymap software, and for the period 1400 to 1300 BCE (Ref. VI.8). Note the following:

1. Astronomical reckoning (designated Universal Time) includes a year zero which is not present in the BC/AD years. It is necessary to ADD one year to the astro-column to convert to calendar years.

2. The distances between the planets are given in A.U. (astronomical units). An A.U. is the average distance between Earth and the sun (150 million kilometers).

3. Most of the closest encounters, at less than 0.4 A.U., occur during the months of July and August.

4. Although we will use Dr. McCluskey's results in the following discussion, we need to recognize some possible limitations on drawing definitive conclusions based on these tables. Most desktop software uses static orbital elements and does not account for long term periodic changes, resulting from small but significant perturbations in the perihelion and eccentricities of the orbits (gravitational effects of Jupiter and the other planets). No two perihelic oppositions are ever exactly the same.

According to Dr. McCluskey's results, there are five oppositions in the examined 100-year period, in which the distances between Earth and Mars are less than 0.4 A.U. (60 million km.). These are listed in Table VI.1.

Table VI.1 – Perihelic Oppositions for the 14th century BCE

Astronomical Year	Calendar Year (BCE)	Date	Separation in A.U.
-1311	1312	Aug. 01	0.3895
-1328	1329	June 12	0.3831
-1343	1344	July 14	0.3774
-1375	1376	June 26	0.3762
-1390	1391	July 28	0.3728

In addition to the 79 year cycle, perihelic oppositions also occur every 15 to 17 years.

The opposition of Aug 1, 1312 BCE is of particular interest here, since it corresponds to the year of the Exodus using the Bible's chronology, where all the years prior to the Revelation at Sinai were solar years. However, the first Passover took place in the springtime and not in the summer. At the time of the Exodus, Mars would be just beginning its transition from an inauspicious star lost in the celestial canopy to the bright blood-red Ra'ah.

NASA has published an evocative simulation which shows the transformation of Mars' image during the recent opposition of 2003, which peaked on August 28. The image of Mars in the opposition of 1312 BCE would have reached maximum size and brightness three months after the Exodus. This three month discrepancy therefore poses a serious challenge to our basic thesis.

We turn, once again, to the Bible for a possible solution to the problem. A favorite and effective tool of Biblical exegesis is the question:" Where else in the Bible do we find the same word or phrase?" Ra'ah shows up in Portion *Ki Thisa* (Exodus XXXII.12), following the sin of the Golden Calf. Moses is informed by God that his people have played the harlot under His wedding canopy, only 40 days after the Revelation at Sinai. Moses pleads and prays against the pending destruction of this errant nation, saying:

"למה יאמרו מצרים לאמר ברעה הוציאם להרג אתם בהרים ולכלתם מעל
פני האדמה שוב מחרון אפך והנחם על הרעה לעמך"

"Why should the Egyptians be able to speak and say: 'for Evil did He bring them out to kill them in the mountains and to wipe them out from the face of the earth.' Withdraw from your fierce anger and refrain from this Evil against thy people"

According to the Bible, these events occurred on the 17th of Tammuz, exactly 90 days after the Exodus (50 days between Passover and the Revelation and 40 days spent by Moses on Mt. Sinai). The

17th of Tammuz is a fast day and begins a three week period of mourning culminating in Tisha b'Av (the 9th of the month of Av), which commemorates the destruction of both the First and Second Temples. This three week period always falls between July and mid-August, so that during the episode of the Golden Calf, Mars would have been at or near its maximum size. The identification of Ra'ah as Mars at the perihelic opposition of August 1, 1312 BCE appears to be quite conclusive, but there is still an open question on the Midrashic interpretation of Pharaoh's warning; did it refer to a star already rising in the wilderness well before the Exodus, or was it a prediction based on the acknowledged prowess of Egyptian astronomy and astrology?

According to one Torah scholar, Pharaoh's warning about Ra'ah was made about two months before the Exodus. The plagues of locusts, darkness and slaying of the first-born were yet to happen. The apparent discrepancy would then increase to five months. Let us revisit Rashi's quotation of the Midrash on Exodus x:10: "I (Pharoh) *see by my astrology* the star (Ra'ah) that signifies blood and slaughter rising toward you in the wilderness." It is quite plausible that Pharaoh and his advisors were indeed making a prediction of the approaching perihelic opposition of Mars, based on more than 1000 years of meticulous astronomical observations and record keeping by the Egyptian civilization.

References – Chapter VI

Ref. VI.1. Rudolph Cohen, "The Excavations at Kadesh Barnea (1976–1978)," *The Biblical Archaeologist*, 44 No.2:93–107 (1981)

Ref. VI.2. Merneptah, the son of Ramses II, became Pharaoh in 1213 BCE. He invaded Canaan about 1210 BCE. The inscriptions on the upright slab, known as the Merneptah Stele, have been translated as follows: "Canaan has been plundered in every evil way. Ashkelon has been brought away captive. Gezer has been seized. Yenoam has been destroyed. Israel is devastated having no seed"

Ref. VI.3. Eliezer Shulman, *The Sequence of Events in the Old Testament*, MOD Publishing House, 1987 (compiled during his exile in Siberia)

Ref. VI.4. James K. Hoffmeier, "Out of Egypt – The Archaeological Context of the Exodus," *Biblical Archaeology Review*, 33 No. 1:30 (2007)

Ref. VI.5. Trudy Dothan and Moshe Dothan, *People of the Sea – the Search for the Philistines*, Macmillan (1992)

Ref. VI.6. John Rummel (Planetary Protection Officer at NASA headquarters), "How Close Mars?" [http://webpages.charter.net/darksky25/Astronomy/Articles/April2003.html]

Ref. VI.7. Jean Meeus, *Astronomical Tables of the Sun, Moon and Planets*, William-Bell (1983) – see also ISBN 0-943396-21-2

Ref.VI.8. Private communication from Dr. McCluskey, May 23, 2006, (*scmcc@wvu.edu*)

Acknowledgements

About two years ago, my sons Michael and Jonathan convinced me to put in writing my speculative interpretations of the bible as viewed from a scientific perspective. This book is the result. During the course of this work, my Israeli grandchildren (Moshe/Momo, Yossi, Itamar and Nina) provided invaluable help and answers on my questions on Gemara, Midrashim and sources thanks to their education at Horev. My three sons (Michael, Benjamin and Harry) together with their wives and children also extended advice and encouragement during my all-to-rare visits to the West Coast of the USA.

My teachers, Rabbi Daniel Lapin and Professor Yosef Bodenheimer have given me inspiration as well as knowledge on the many aspects of the harmony between science and the sacred texts.

I am grateful to Professor Dina Ben Yehudah, Head of the Hematology Department at Hadassah Ein Kerem and her wonderful staff who have maintained me in good health despite some serious episodes some four years ago.

I wish to dedicate this work to my wife, the mother of our four sons, Renate Rosa Hirsch Medved, z.l.

About the Author

Dr. David Medved began his career in science as a control chemist in an alcohol/acetone plant at the age of 16, working fulltime while completing his studies at Central H.S. in Philadelphia. He was awarded a Mayor's Scholarship just in time for his enlistment in the US Navy as a Radar Tech. Upon his discharge, he returned to the University of Pennsylvania as a chemistry major and went on to a Masters and PhD in Physics while working as a research engineer at Philco. Upon receiving his PhD, Dr. Medved left Philadelphia for San Diego where he joined the Convair Division of General Dynamics. Upon the launch of Sputnik, he was appointed as a Group Leader of 10 other PhD specialists working on "Physics Research Against Ballistic Missiles" some 20 years before President Reagan's Star Wars. The team's unclassified publications appeared in the *Physical Review*, *Advances in Electronics* and the *Journal of Applied Physics*. During much of this period, Dr. Medved was teaching solid state physics and physical electronics at UCLA Extension and at San Diego State College. He was selected by NASA to serve as a Principal Investigator on the Gemini project which brought him into close contact with the second generation of the

Astronaut corps. As the PI on the project he was responsible for design of the ion and electron particle detectors to be mounted on the Agena satellite and planning the trajectory of the Gemini spacecraft as it maneuvered toward rendezvous and docking (a dress rehearsal for the Apollo moon mission). Later, Dr. Medved decided to become a scientist-entrepreneur, forming MERET Inc., pioneering the design, fabrication and installation of short range fiber optic communication systems. MERET is still an operating company 35 years after its founding. Following the sale of MERET to AMOCO, Dr. Medved immigrated to Israel where he started JOLT Ltd. (Jerusalem Optical Link Technologies) on the campus of JCT. JOLT is a pioneer in wireless optical communications (FSO) and was acquired by MRV Communications in May 2000. Dr. Medved currently serves as the Chief Technical Officer of MRV Jerusalem, in addition to his service with JCT.